Lecture Notes
in Business Information Processing 239

More information about this series at http://www.springer.com/series/7911

Heiner Stuckenschmidt · Dietmar Jannach (Eds.)

E-Commerce and Web Technologies

16th International Conference
on Electronic Commerce and Web Technologies, EC-Web 2015
Valencia, Spain, September 2015
Revised Selected Papers

 Springer

Editors
Heiner Stuckenschmidt
Institut für Informatik, B6.26
University of Mannheim
Mannheim
Germany

Dietmar Jannach
Department of Computer Science
TU Dortmund
Dortmund
Germany

ISSN 1865-1348   ISSN 1865-1356   (electronic)
Lecture Notes in Business Information Processing
ISBN 978-3-319-27728-8   ISBN 978-3-319-27729-5   (eBook)
DOI 10.1007/978-3-319-27729-5

Library of Congress Control Number: 2015957227

Printed on acid-free paper

This Springer imprint is published by SpringerNature
The registered company is Springer International Publishing AG Switzerland

Preface

EC-Web is an international scientific conference series devoted to technology-related aspects of e-commerce and e-business. The 16th edition of the conference, EC-Web 2015, took place in Valencia, Spain, in September 2015 and served as a forum to bring together researchers and practitioners to present and discuss recent advances in their fields. The conference series historically covers the following areas:

- Search, comparison and recommender systems
- Preference representation and reasoning
- Semantic-based systems, ontologies and linked data
- Agent-based systems, negotiation and auctions
- Social Web and social media in e-commerce
- Computational advertising
- E-commerce infrastructures and cloud-based services
- Service modelling and engineering
- Business processes, Web services and service-oriented architectures
- E-business architectures
- Emerging business models, software as a service, mobile services
- Security, privacy and trust
- Case studies

This year, the conference program focused on two main topics, recommender systems and matchmaking as well as social and Semantic Web aspects of electronic commerce. The works presented at the conference reflect recent trends in different subfields related to e-commerce and Web technologies, which can be summarized as follows.

In the Web era, resource retrieval techniques play a fundamental role in helping users deal with the issues related to information overload. Today's search engines and matchmaking systems no longer work solely on plain text documents and keyword analysis. Their results are computed and enriched by putting together various types of information, e.g., link-based data, user preferences, as well as information encoded in so-called knowledge graphs. The same happens for personalized information filtering systems, in particular for recommender systems, where collaborative and content-based approaches are combined to exploit diverse data sources in order to improve not only the accuracy of the results but also their novelty, diversity, and serendipity.

At the same time, the Social Web has become an important platform for activities related to e-commerce on the Web. Social networks, review sites, and blogs have become important places to market products and analyze their reception by larger groups of customers. Thus, methods for creating and analyzing the behavior of users on the Social Web become more and more important. Recent topics of research in that direction include targeted advertisement, opinion mining and sentiment analysis, as well as trust and reputation mechanisms in social online social networks and their impact on user behavior.

Finally, Semantic Web technologies, in particular data markup standards today provide an established means for publishing and partially exchanging structured data on the Web. Large numbers of websites have started to markup their content using standards such as Microdata, Microformats, and RDFa allowing search engines like Google, Bing, and Yahoo! to use this markup to improve their search results. At the same time, such approaches are increasingly used in the e-commerce area where structured product descriptions are published online and sometimes even linked to general purpose schemas such as schema.org and or product classifications such as eClass.

Overall, we received 28 paper submissions for the conference, which addressed a variety of these topics. Each submission was reviewed by three members of the Program Committee. Based on the judgement of the reviewers, ten papers where selected for publication in these proceedings. This corresponds to an acceptance rate of 35 %.

The section on recommender systems in the proceedings contains three papers: Peska and Vojtas investigate the use of implicit user feedback in terms of interactions with a shop website to determine preference relations between products. They argue that this technology is particularly useful for small retailers that do not have a sufficiently large number of sales to use conventional methods. Kaminskas et al. also investigate the problem of providing recommender solutions for small retailers. They report experiences from practical applications in different companies and present different approaches to overcome the data sparsity problem. Ristoski et al. finally investigate the use of freely available data on the Web to create a better basis for comparing different products.

The three papers in the second section of the proceedings address the problem of recommending multimedia content. Hatem et al. propose an extension of LDA-based approaches for image retrieval. The apply the classic LDA model on the textual context of images to determine associated topics that are then compared with the topics of user queries to retrieve images. They show that comparing topics rather than words improves the retrieval performance. Cremonesi et al. propose an approach for the personalized recommendation of TV channels. Their results indicate that the viewing times are the most accurate predictors of user interests. Deldjoo et al. finally investigate the use of low-level image features to improve video recommendation. They show that low-level features are not only a solid basis for recommendation, but can also help to outperform approaches that use high-level annotations in terms of the genre.

The papers in the Semantic and Social Web section of the proceedings investigate the use of publicly available data for building applications. Meusel et al. present an analysis related to structured product information available on the Web. Their work shows that an impressive amount of product information is available on the Web. However, despite numerous standardization efforts, heterogeneity is still a major problem. Zhou and Guo present an empirical study regarding the helpfulness of product reviews in social media. They show that there is a combined effect of text sentiment and rating that shows a better correlation to helpfulness than when considering the variables in isolation. Santos et al. finally investigate the use of Twitter data to derive insights on the Brazilian stock market and compare their insights with previous research on Twitter data from English-speaking users.

Finally, in the last paper of the proceedings, Mousheimish et al. provide additional arguments that prediction and proactive replanning can improve logistic processes.

November 2015   Heiner Stuckenschmidt
                                                                Dietmar Jannach

Organization

EC-Web 2015 was organized in the context of the 26th International Conference on Database and Expert Systems Applications (DEXA 2015) and took place during September 3–4, 2015, in Valencia.

Conference Organization

Co-chairs

Heiner Stuckenschmidt   University of Mannheim, Germany
Dietmar Jannach   TU Dortmund, Germany

Senior Program Committee

Jorge Cardoso   University of Coimbra, Portugal
Tommaso Di Noia   Politecnico di Bari, Italy
Conor Hayes   National University of Ireland, Galway
Terry R. Payne   University of Liverpool, UK

Program Committee

Ardissono, Liliana	University of Turin, Italy
Aumayr, Erik	NUI Galway, Ireland
Bae, Joonsoo	Chonbuk National University, South Korea
Basile, Pierpaolo	University of Bari Aldo Moro, Italy
Bellogín, Alejandro	Universidad Autonoma de Madrid, Spain
Breslin, John	National University of Ireland Galway, Ireland
Bridge, Derek	University College Cork, Ireland
Burke, Robin	DePaul University, USA
Camarinha-Matos, Luis M.	Universidade Nova de Lisboa, Portugal
Cena, Federica	University of Turin, Italy
Chapman, Archie	University of Sydney, Australia
Cremonesi, Paolo	Politecnico di Milano, Italy
Cuzzocrea, Alfredo	ICAR-CNR and University of Calabria, Italy
Dabrowski, Maciej	Altocloud.com, Ireland
de Gemmis, Marco	University of Bari, Aldo Moro, Italy
De Luca, Ernesto William	FH Potsdam, Germany
Düdder, Boris	TU Dortmund, Germany
Eckert, Kai	Hochschule der Medien Stuttgart, Germany
Felfernig, Alexander	Graz University of Technology, Austria
Ferreira Da Silva, Catarina	University of Lyon 1, France

Contents

Recommender Systems

Using Implicit Preference Relations to Improve Content Based
Recommending.   3
   Ladislav Peska and Peter Vojtas

Product Recommendation for Small-Scale Retailers.   17
   Marius Kaminskas, Derek Bridge, Franclin Foping, and Donogh Roche

Using Graph Metrics for Linked Open Data Enabled Recommender
Systems.   30
   Petar Ristoski, Michael Schuhmacher, and Heiko Paulheim

Multimedia Recommendation

Toward Building a Content-Based Video Recommendation System Based
on Low-Level Features   45
   Yashar Deldjoo, Mehdi Elahi, Massimo Quadrana,
   and Paolo Cremonesi

Personalized and Context-Aware TV Program Recommendations Based on
Implicit Feedback   57
   Paolo Cremonesi, Primo Modica, Roberto Pagano, Emanuele Rabosio,
   and Letizia Tanca

An LDA Topic Model Adaptation for Context-Based Image Retrieval   69
   Hatem Aouadi, Mouna Torjmen Khemakhem, and Maher Ben Jemaa

Social and Semantic Web

Exploiting Microdata Annotations to Consistently Categorize Product
Offers at Web Scale   83
   Robert Meusel, Anna Primpeli, Christian Meilicke, Heiko Paulheim,
   and Christian Bizer

The Interactive Effect of Review Rating and Text Sentiment on Review
Helpfulness   100
   Shasha Zhou and Bin Guo

A Twitter View of the Brazilian Stock Exchange Market   112
   Hugo S. Santos, Alberto H.F. Laender, and Adriano C.M. Pereira

Process Management

Towards Smart Logistics Process Management   127
   Raef Mousheimish, Yehia Taher, and Béatrice Finance

Author Index   139

Recommender Systems

Using Implicit Preference Relations to Improve Content Based Recommending

Ladislav Peska[✉] and Peter Vojtas

Faculty of Mathematics and Physics Charles, University in Prague,
Malostranske namesti 25, Prague, Czech Republic
{peska,vojtas}@ksi.mff.cuni.cz

Abstract. Our work is generally focused on recommending for small or medium-sized e-commerce portals, where we are facing scarcity of explicit feedback, low user loyalty, short visit times or low number of visited objects. In this paper, we present a novel approach to use specific user behavior as implicit feedback, forming binary relations between objects. Our hypothesis is that if user select some object from the list of displayed objects, it is an expression of his/her binary preference between selected and other shown objects. These relations are expanded based on content-based similarity of objects forming partial ordering of objects. Using these relations, it is possible to alter any list of recommended objects or create one from scratch.

We have conducted several off-line experiments with real user data from a Czech e-commerce site with keyword based VSM and SimCat recommenders. Experiments confirmed competitiveness of our method, however on-line A/B testing should be conducted in the future work.

Keywords: Content-based recommender system · Implicit preference relations · VSM · User preference · E-Commerce

1 Introduction and Related Work

Recommending on the web is both an important commercial application and a popular research topic. The amount of data on the web grows continuously and it is virtually impossible to process it directly by a human. Also the number of trading activities on the web steadily increases for several years. Various tools ranging from keyword search engines to e-shop parameter search were adopted to decrease information overload. Although such tools are definitely useful, users must specify in detail what they want.

Recommender systems are complementary to this scenario as they are mostly focused on serendipity – finding surprising, unknown, but interesting items and presenting them to the user. Our aim is to enhance recommending systems on e-commerce domains, more specifically on small to medium enterprises without dominant position on the market. This brings some unique challenges such as unwillingness of user to register, scarcity of explicit feedback [7], low consumption rate, user loyalty, number of visited objects [17] etc. Thus we are focusing on learning from implicit user feedback, algorithms with fast learning curve, using additional content information as well as improving methods for mining and interpreting user behavior.

© Springer International Publishing Switzerland 2015
H. Stuckenschmidt and D. Jannach (Eds.): EC-Web 2015, LNBIP 239, pp. 3–16, 2016.
DOI: 10.1007/978-3-319-27729-5_1

1.1 Motivation of Current Research

During his/her visit, user often has to evaluate multiple objects at once. Either on catalogue pages, after a search query or as recommended item – a list of objects is presented to the user. The user typically examines (some of) them and eventually opens detail of an object (or more objects). We interpret this user behavior as follows:

- User put some effort to evaluate objects (the effort is indicated by user behavior e.g. mouse over, scrolling, visible time etc.).
- If user selects (i.e. click on) some objects and opens their details, these objects more likely correspond with his/her preference than the others (visible, but ignored). According to the hypothesis above, we can create partial preference ordering
- IPR_{rel}: Pref(Visible & clicked) > Pref(Visible & ignored).
- However we cannot say anything about objects which were not examined at all (e.g. because they were at the bottom of the list and thus not visible). The user could have opened them if he/she had noticed them.

Our approach is illustrated on a sample e-commerce website on Fig. 1.

The IPR_{rel} relation itself has very limited prediction capability as it only describes past user behavior. Thus it needs to be expanded e.g. along the content-based similarity of the objects.

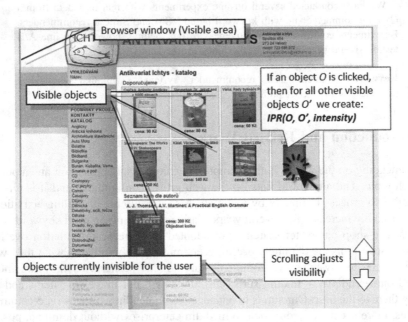

Fig. 1. Illustrative example of our approach. On a sample website (category page of a secondhand bookshop) are displayed several objects. Some of them are within visible area, some not, depending on the browser properties. Visible area can be shifted by scrolling. The click on object thus can be considered as a behavior favoring this object over other visible ones.

1.2   Main Contributions

The main contributions of this paper are:

- Novel content-based recommending method based on Implicit Preference Relations (IPR) integrating areas of content-based recommending, implicit feedback mining and preference relations.
- Experiments on real users from Czech secondhand bookshop.
- Datasets of complex user behavior for future experiments. The observation is ongoing, so the datasets are extended continuously.

1.3   Related Work

Implicit Feedback Interpretation: Contrary to the explicit feedback, usage of implicit feedback requires no additional effort from the user of the system. Monitoring implicit feedback varies from simple user visit or play counts to more sophisticated ones like scrolling or mouse movement tracking [10, 21]. Due to its effortlessness, data are obtained in much larger quantities for each user. On the other hand, data are inherently noisy, messy and hard to interpret [8].

Our work lies a bit further from the mainstream of the implicit feedback research. To our best knowledge, the vast majority of researchers focus on interpreting single type of implicit feedback [3], proposing various latent factor models [8, 20], its adjustments [7] or focusing on other aspects of recommendations using implicit feedback based datasets [1, 19]. Also papers using binary implicit feedback derived from explicit user rating are quite common [13].

We are generally more interested in defining and interpreting novel types of implicit user behavior and modelling user's preference based on multiple types of implicit feedback. We can trace such efforts also in the literature. One of the first paper mentioning implicit feedback was Claypool et al. [2] comparing several implicit preference indicators against explicit user rating. This paper was our original motivation to collect and analyze various types of user behavior to estimate user preference [15]. More recently Yang et al. [21] analyzed several types of user behavior on YouTube. Authors described both positive and negative implicit indicators of preference and proposed linear model to combine them. Also Lai et al. [10] work on RSS feed recommender utilizes multiple reading-related user actions. However the lack of publicly available datasets containing complex user behavior hinders future development of the area. This was our motivation to propose the IPIget JavaScript component [14] for mining complex user behavior from unregistered e-commerce users and also to publish the dataset used during our experiments. One of the interesting challenges of implicit feedback is deriving negative preference. Various approaches were applied ranging from adding negative preference with small weight to all the unvisited objects [8], considering low amount of implicit feedback as negative preference [16] or considering some user behavior as negative preference [11, 21]. From this point of view, our current approach is somewhat similar to [11], as we create binary relations of more/less preferred objects based on users observe & ignore and observe & open behavior.

Our work can be also viewed as context-aware recommendation, where other objects displayed on the page serve as context to the event of selecting one of them. Within this point of view, our work is similar to Eckhardt et al. [5], suggesting that user rating should be considered in the context of the list of displayed objects.

Preference Relations: In the area of preference relations, methods focus on comparing two objects, forming partial ordering $O_A >_{rel} O_B$. We would like to mention two papers proposing recommender system based on preference relations. Yi Fang et al. [6] used click-through data from nano-HUB and proposed a latent pairwise preference learning approach. Deskar et al. [4] proposed matrix factorization based on preference relations from explicit user feedback. However both authors proposed CF algorithms to utilize preference relations, which is not very suitable for small e-commerce portals. As our previous work [17, 18] suggests, CF methods cannot predict well under the constraints of continuous cold start problem affecting small e-commerce portals. Thus we focused on content-based (CB) and hybrid algorithms which cope better with cold start problem. Yi Fang et al. [6] also brought an interesting idea that the position of object in the list affects the likelihood of being clicked by the user (the bottom of the list is often not evaluated at all). In our work, we developed this observation further as we can detect not just its position, but also how much (e.g. in terms of visible time, mouse over count etc.) was each object examined.

2   Implicit Preference Relations (IPR)

In this section, we will briefly describe our model of collecting implicit preference relations (IPR). For each user session and each visited page, we use IPIget component to collect implicit user feedback. Part of this feedback is also location of objects (products) within the page, information about browser window size and position of viewed page in the browser window over time. From these information we can recreate whether each object was visible for the user, for how long and e.g. on which part of the visible window. Moreover we collect information about user evaluation of the objects e.g. that user clicked on some of them. The information is transformed into two relations:

   Visible(PageVisitID, OID, Clicked, VisibleTime) and
   Visit(PageVisitID, TotalTime), where:

- *PageVisitID* is an ID of single visit of a webpage by current user.
- *OID* is an ID of object displayed on the webpage identified by *PageVisitID*.
- *Clicked* is binary information whether user opened the detail of the object.
- *VisibleTime* is number of seconds, when object was present at visible area.

- *TotalTime* is total number of seconds, the user spent on the page.
- *VisibleRel* is relative visit time defined as *VisibleTime/TotalTime* for each pair of *OID, PageVisitID*.

For each user U, based on his/her *Visible* relation, we define $IPR_{rel}(oid_1, oid_2, intensity)$ relation describing the user behavior as follows: if oid_1 was selected by the user (e.g. clicked) and oid_2 was visible, but ignored, then oid_1 is more preferred than oid_2 with some *intensity*. The current implementation of our method defines intensity as follows:

$$intensity \sim\ = min\left(\frac{vis(oid_2)}{vis(oid_1)}, 1\right) \tag{1}$$

$$vis(oid) \sim\ = nVisibleAbs + VisibleRel - (nVisibleAbs * VisibleRel) \tag{2}$$

Idea behind comparing visibilities of objects (1) is that if the difference between visibilities of the objects is high, then maybe user did not select oid_2 just because he/she did not noticed it. *Vis(oid)* (2) is defined as *probabilistic sum*[1] of normalized *nVisibleAbs* and *VisibleRel* values of the current row. Due to highly skewed upper bound of *VisibleTime*, we decided to use $Q_{0.9}(VisibleAll)$ linear normalization to the 90 % quantile of *VisibleTime* over all users to omit outliers (3). We use probabilistic sum instead of e.g. average or max as we expect some mutual benefit, if both *nVisibleAbs* and *VisibleRel* values are high. Other fuzzy-logic disjunctions could be used too, but as this is not the key part of the paper, we opted for using a simple one like this.

$$nVisibleAbs = min\left(\frac{VisibleTime}{Q_{0.9}(VisibleAll)}, 1\right) \tag{3}$$

MinVisibility threshold defines minimal necessary visibility *vis(oid)* to create IPR_{rel} relation. If more than one object is clicked, the relation between them is neutral. IPR_{rel} itself can be used e.g. to filter out uninteresting objects from the search results etc., but its prediction capability is low. Hence we define $\widehat{IPR}_{rel}$ increasing extension of original IPR_{rel} based on assumption that similar objects to oid_1, and oid_2 will be evaluated similarly by the user. For each IPR_{rel} (O_1, O_2, $int_{1,2}$), L_1 and L_2 are lists of objects similar to O_1 and O_2 respectively. The ***minSimilarity*** threshold applies in order to qualify into each list. Note that object similarities can be precomputed (the object attributes are relatively stable over time) and only the ones with higher-than-threshold similarity have to be stored. In current implementation, the cosine similarity over object content-based attributes with TF-IDF weighting is computed. However as the choice of similarity method is completely orthogonal to the other components, it can be easily

---

[1] Probabilistic sum $S_{sum}(a, b) = a + b - a * b$.

changed in the future, even for some CF based method. Having $O_x \in L_1$ and $O_y \in L_2$, the intensity of $\widehat{IPR}_{rel}$ (O_X, O_Y, $int_{x,y}$) based on $IPR_{rel}(O_1, O_2, int_{1,2})$ is defined as:

$$int_{x,y} \sim = int_{1,2} * sim(O_X, O_1) * sim(O_Y, O_2) \qquad (4)$$

Resulting intensity $int_{x,y}$ is a sum of intensities of all possible (with sufficient similarity) derived relations from some $IPR_{rel}(O_1, O_2)$. Note that if relation between objects is inverted, the intensity is subtracted. The **minIntensity** threshold defines minimal intensity of $\widehat{IPR}_{rel}$ relation in order to be retained.

3   Recommending Algorithms for Implicit Preference Relations

3.1   From Partial Ordering to Ranked List

For an arbitrary fixed user, $\widehat{IPR}_{rel}$ forms partial ordering, where many pairs of objects are incomparable. Due to *minSimilarity* and *minIntensity* thresholds, there might not be any evidence about user preference on similar objects. Nevertheless, we may have a linearly ordered preference list *objList* for this specific user from other recommenders. So, one of the important tasks is to merge new preference partial ordering with the previous ordered list of objects into a new ranked list of objects. Our approach is based on the *intensity* parameter of $\widehat{IPR}_{rel}(O_1, O_2, intensity)$ and formed by the following requirements:

1. The higher the *intensity* is, the better evidence about user preference we have. Thus the distance between O_1 and O_2 in the ranked list should be also larger.
2. If a subset of relations between objects forms a circle, the object with the most intense positive relation should appear highest in the list of objects. The relations with higher intensity have priority over the ones with lower intensity.
3. Merging two orderings a conflict may arise. Suppose R is partially constructed ranked list: $O_1, O_2 \in R$ and *Position* (O_1) > *Position* (O_2) (i.e. O_2 is more preferred than O_1). Suppose we have relation in the opposite direction $\widehat{IPR}_{rel}(O_1, O_2, int)$ with some intensity *int*. We should re-rank objects to cope with the new relation only if its intensity is high enough and also does not violate too many existing relations in terms of relative distance between those two object in the list ordered by previous observations.

The Algorithm 1 describes transformation of $\widehat{IPR}_{rel}$ into the ranked list of objects. Note that an initial list of objects from other recommender can be passed as input and then re-ranked according to the $\widehat{IPR}_{rel}$ relations. Otherwise the algorithm will build the list from scratch. We designed 3 variants of coping with re-ranking of objects with following semantics: If our main intention is to filter out uninteresting objects, the best way is to push *back* objects which are inferior to some others. If we don't want to miss potentially interesting objects, we should use prioritizing variant and bring *forward* objects with positive relation. The *swap* variant utmost take into account the requirement 1 by maximizing distance between objects with high-intensity relations. Please note that relations are considered according the increasing order of their intensity (thus more intense relations can override less intense) and re-ranking is only performed if ratio between relation *intensity* and *relative distance* is high enough. The relative distance check was added after some preliminary experiments as we wanted to avoid situations such as that original ordering received from some base recommender was completely overridden by even a very weak relations.

Algorithm 1: Algorithm IPR-rank for merging $\widehat{IPR}_{rel}$ and *objList* to the ranked list of objects. IPR-rank access ordered list of $\widehat{IPR}_{rel}$ relations starting from smallest intensity and then constructs or update the *objList* as follows: If the objects of the relation are not yet in the list, they are added to the head and tail resp. to maximize their distance. If they are both in the list, but in the opposite order, the *relative distance* between them (with respect to the total number of objects) is calculated. If the *intensity* of the relation is higher than the *relative distance*, the objects are either *swapped*, the better is *moved forward* or the worse is *moved backward* based on the current variant of the algorithm. The relations with the higher *intensity* are considered later so they are less likely to be changed. See illustrative example on Fig. 2.

```
function IPR-rank(IPR̂ᵣₑₗ, conflictStrategy, objList){
    /*IPR̂ᵣₑₗ is ordered from lower to higher intensity */
    foreach((o1, o2, int) ∈ IPR̂ᵣₑₗ){ /*i.e. o1 is clicked and o2 ignored*/
        if([o1,position1],[o2,position2] ∈ objList){
            if(position1 > position2){ /*i.e. there is a conflict*/{
                relDist = (position2 - position2)/Count(objects)
                if(int > relDist){ /*i.e. the relation intensity is significant*/
                    switch(conflictStrategy){
                        case: forward
                            Object o1 is moved just before o2
                        case: backward
                            Object o2 is moved just behind o1
                        case: swap
                            Position of objects o1, o2 is swapped
            } } }
        }else{
            if(o1∉objList){objList = o1 CONCAT objList;}
            if(o2∉objList){objList = objList CONCAT o2;}
} } }
```

3.2   Combining IPR-Rank with Other Recommending Algorithms

One of the disadvantages of sole IPR-rank algorithm is that it can't rank all objects as well as we might not have any relations for some users. It is more natural for our approach to be combined with some other method capable to provide full list of ranked objects for each user. Therefore we implemented three recommending algorithms serving as both baselines and initial ranked list for the IPR-rank algorithm.

Vector Space Model (VSM) is well-known content-based algorithm brought from information retrieval. We use the variant described in [12] with binarized content-based attributes serving as document vector, TF-IDF weighting and cosine similarity as objects similarity measure. No adjustments towards increasing diversity or novelty were performed as those metrics were not evaluated in the experiments. The algorithm recommends top-k objects most similar to the user profile.

Stochastic Gradient Descent Matrix Factorization (SGD MF). Using vector of latent factors common for both users and objects followed by some matrix factorization technique is currently state of the art technique for collaborative filtering. Based on our previous experiments [17], we discourage using collaborative techniques on small e-commerce due to the low volume of user interactions. However we implemented SGD MF according to [9] to compare it against content-based techniques and to measure improvement of hybrid approach by combining it with IPR.

SimCat **Hybrid Recommender.** The last recommender, named *SimCat*, is a simple hybrid approach based on collaborative similarity of product categories. It is motivated by the problem of too shrink categories, which sometimes contains too few objects to be presented to the user. So if the user is in such a shrink category, it might be useful to recommend him/her also objects from categories similar to the current one. Also as the number of categories is less than the number of objects by the order of magnitude (and the list of categories is much more stable), it is possible to use collaborative-based similarity of categories also in our scenario. *SimCat* first computes lists similar categories $L_c < C_i$, $Sim(C, C_i) >$ for each product category C based on users co-visiting either categories or products from these categories. Suppose U_{C1} and U_{C2} are sets of users who visited category C_1 and C_2 respectively. Then the similarity of categories $Sim(C_1, C_2)$ is defined as Jaccard similarity of the user sets (5):

$$Sim(C_1, C_2) \sim = \frac{|U_{C1} \cap U_{C2}|}{|U_{C1} \cup U_{C2}|} \tag{5}$$

The list of recommendations is calculated as follows. For arbitrary fixed user U, the frequency of his/her visits to all categories is listed. Let $L^U < C_i, F_i >$ is the list of categories C_i with non-zero frequency F_i. The list is then expanded according to the L_{Ci} for each category C_i. Recommended objects are ordered according to the corresponding category and then randomly. See Algorithm 2 for more details.

Algorithm 2: Computing list of recommendations with *SimCat* algorithm. For user U and list of his/her visited categories L^U, the method computes expanded list of categories $\widehat{L^U}$, which contains original L^U and all categories similar to ones in L^U. The score of category C_j in $\widehat{L^U}$ is defined as the sum of all similarities Sim_j to some $C_i \in L^U$ multiplied by its frequency F_i. Each object O_k receives score of its corresponding category C_l plus small random ε.

```
function SimCat(U, L^U){
  L̂^U = L^U;
  foreach(<C_i,F_i> ∈ L^U){
    foreach(<C_j,Sim_j> ∈ L_Ci)
      if(<C_j,_> ∉ L̂^U){
        Add <C_j, Sim_j* F_i > to L̂^U;
      }else{/*suppose <C_j,S_j>∈ L̂^U*/
        Alter value of S_j: S_j += Sim_j* F_i;
  } } }
  foreach(object O_k: O_k belongs to category C_l){
    score(O_k) = S_l + ε;
  }
  Order objects according to its score; return objects;
}
```

We suppose that *SimCat* algorithm would result poorer than *VSM* as it comprise a lot of randomness, however it might be a good alternative to e.g. showing only most popular objects etc. In this setting we focused on categories of the products, as it is natural way to present and filter our datasets (e.g. catalogue pages), but virtually any content-based attribute or its combination can be used. Another reason for using *SimCat* was evaluation, how much can *IPR-rank* improve over relatively weak recommender.

3.3   Implicit Preference Relations - Example

In this subsection, we would like to illustrate our approach on a small example (see Fig. 2). Suppose we have a catalogue page with objects O1– O8 (Fig. 2a). For time T1 user observes top of the page and after a while, he/she scrolls a bit lower (T2) and opens detail of O3. Objects O7 and O8 was outside of the visible area. IPR_{rel} relations are mined from user behavior according to (1), (2), (3) (Fig. 2b). The IPR_{rel} are enlarged via content-based similarity into $\widehat{IPR}_{rel}$ according to (4). This relations are then processed with IPR-rank algorithm (Algorithm 1) together with original list of recommended objects (Fig. 2c), forming an enriched recommended list (Fig. 2d). Steps of *forward* and *backward* variants of Algorithm 1 are shown on right (Fig. 2e).

Note that updated objects for each step are in bold. The relations are calculated assuming T1 = 15 s, T2 = 5 s and $Q_{0.9}(VisibleAll) = 100$ s.

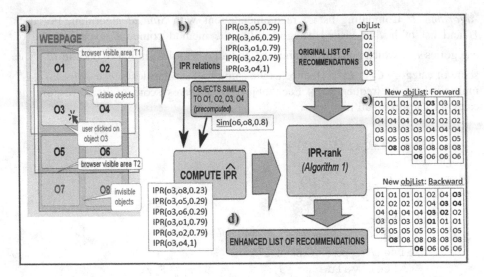

Fig. 2. Illustrative example of creating $\widehat{IPR}_{rel}$ relations from user behavior.

4   Experiments

4.1   Experimental Datasets

In this section we provide details about series of off-line experiments with several variants of IPR-rank algorithm. The experiments were conducted on Czech secondhand bookshop called Antikvariat Ichtys (www.antikvariat-ichtys.cz). The bookshop dataset contains several attributes for each object: *book title, author name, book category, publisher, publishing date, short textual description* and *book price*. The attributes were binarized and used by VSM recommender. The dataset contains user behavior observed during the period of February to early December 2014. The IPIget component collected data from in total approx. 22000 users visiting 40000 pages, however only some of users used catalogue pages (5500) and only a fraction of them could be observed over a sufficiently long period of time to create train and test set (1760 users). We demanded at least 3 visited pages and at least one visited object in the last third of the user data (see Sect. 4.2 for details).

We would also like to mention some special features of the secondhand bookshop domain. Firstly, the quantity of objects (books) in stock is usually only one (i.e. if a user purchase a book, nobody else can buy the same one). This directly implies high object fluctuation. Furthermore, there is relatively high ratio between active objects (approx. 10000) and users (roughly 50–100 unique users per day). Last but not least, majority of the traffic comes from a search engine, where user searches for a specific book. Large portion of the users lands on a specific object and often do not navigate through the website at all. These circumstances makes collaborative filtering methods less relevant as the user-object matrix is simply too sparse and the list of objects changes a lot over time.

4.2   Evaluation Procedure and Success Metrics

The evaluation procedure was carried out as follows: For each user, his/her feedback was divided into two parts (train set, test set) according to the timestamp – two thirds of earlier data formed train set and following one third leaves as test set. Afterwards, for each user, recommending methods produced ranked list of objects which were evaluated according to the success metrics.

The selection of success metrics was a bit tough as the dataset do not contain any explicit feedback. Basically there are two options – either to use purchase event as an evidence of positive preference, or settle for simple viewing object detail. Due to insufficient amount of purchase actions, we opted for viewing detail of an object. More formally, relevance $r_{u,o}$ of object o for user u is defined as:

$$r_{u,o} \sim\ = 1\ IFF\ u\ visited\ o, 0\ OTHERWISE \tag{6}$$

Furthermore we focus solely on ranking metrics as the bookshop's interface does not contain capability to rate books and also objects are presented to the user as ranked list. We adopt normalized distributed cumulative gain ($nDCG$) as a standard metrics to rate relevance of objects list. The premise of $nDCG$ is that relevant documents appeared low in the recommended list should be penalized (logarithmical penalty applied) as they are less likely to attract user attention. This fits well into the recommending scenario, where lower-ranked objects are presented to the user on less desirable positions. Results of overall $recall@top$-k (or shortened as $r@topk$) metric is shown as it has more intuitive nature. We do not consider precision at this scenario, because according to our observations, as long as some relevant items are presented and the list of recommended items is relatively short, the presence of irrelevant items does not substantially affect the human evaluation of recommender systems. Furthermore the effect of precision is further suppressed by using fixed top-k sizes.

4.3   Results and Discussion

In this subsection we would like to present results from the experiments and provide some insight on the IPR-rank method and its hyperparameters. Due to its complexity, we did not conduct full hyperparameter grid search, however each hyperparameter was tested with selected set of other parameter's values. For each hyperparameter, the results are averaged over all settings of other parameters. Table 1 depicts overall results.

As can be seen from the results, VSM based algorithms greatly outperformed SimCat based algorithms as well as sole IPR-rank and Stochastic Gradient Descend Matrix factorization. In case of both SimCat and SGD MF, each tested setting of IPR-rank improved list of recommendations in terms of our success metrics. Higher improvements generally was achieved while using less restrictive thresholds (*minVisibility, minSimilarity, minIntensity*). IPR-rank also improved VSM based recommendations, although the improvement was smaller, however still with some significance (p-value = 0.088). Note that there were no $\widehat{IPR}_{rel}$ relations for some users (they did not visit any category page), which also affected the difference in results. Higher *minSimilarity* threshold improved VSM + IPR-rank method, with best results between

Table 1. Results of the IPR-rank methods in terms of nDCG and recall@top-k: parameters in the brackets are as follows: *minSimilarity, minVisibility, minIntensity*, conflict resolving variant. The IPR-rank parameter settings with best achieved results are displayed. SimCat, VSM, SGD MF and Random ordering represent our baselines.

Method	nDCG	r@5	r@10	r@50
best VSM + IPR-rank (0.5, 0.1, 0.1, swap)	0.475	13.6%	15.7%	20.7%
VSM	0.464	13.2%	15.1%	19.6%
IPR-rank (empty *objList*) (0.5, 0.1, 0.1, swap)	0.247	7.1%	7.7%	8.5%
best SimCat + IPR-rank (0.01, 0, 0.1, forward)	0.219	4.7%	6.3%	10.0%
best SGD MF + IPR-rank (0.01,0.1,0.1, forward)	0.191	3.3%	4.7%	8.2%
SimCat	0.136	0.9%	1.5%	5.4%
SGD MF (500 lat. factors, max 500 iterations)	0.126	0.89%	1.2%	3.3%
Random recommendations	0.085	0.09%	0.14%	0.27%

0.5 and 0.8. This is probably caused by better supplied initial *objList*, making majority of weak relations unnecessary or even incorrect. The poor results of SGD MF supports our previous observations that purely collaborative filtering techniques cannot perform well in domains with such low amount of feedback per user.

The *minSimilarity* threshold seems to be the most important IPR-rank parameter. It indirectly affects total number of relations as well as their average strength. The *minIntensity* threshold would have similar effect, but as it is applied later on, it is highly dependent on *minSimilarity* value. The *minVisibility* threshold have lower impact on the results. As for the variants of IPR-rank conflict solving, the *backward* variant was almost consistently inferior to the other two, perhaps because success metrics are oriented solely on positively preferred objects. The *swap* option is generally more successful if combined with VSM, *forward* with SimCat (see Table 2 for more details).

Table 2. Average values of nDCG for IPR-rank parameters *minSimilarity* and conflict resolving for either SimCat or VSM recommender systems.

MinSimilarity threshold, SimCat+IPR			MinSimilarity threshold, VSM+IPR			
0.01	0.2	0.5	0.2	0.3	0.5	0.8
0.207	0.166	0.154	0.465	0.470	0.473	0.472

Conflict resolving, SimCat+IPR			Conflict resolving, VSM+IPR		
Forward	Backward	Swap	Forward	Backward	Swap
0.173	0.140	0.168	0.465	0.460	0.466

5   Conclusions and Future Work

In this paper, we proposed novel application of specific implicit feedback in the domain of small to medium e-commerce sites. Observing user's behavioral patterns on catalogue pages lets us create preference relations $\widehat{IPR}_{rel}$ between selected and ignored objects. We also proposed approach to extend this relations to similar objects and algorithm IPR-rank to merge them with ordered list of objects from other recommenders. Proposed methods can be viewed also as inferring negative implicit preference or context-aware recommender system (with other objects on page serving as context). Experiments held on Czech secondhand bookshop dataset shown that IPR-rank can improve recommendations coming from other recommending algorithms, although the parameters of IPR-rank should be carefully set. We are currently working on larger scale experiments involving datasets from other e-commerce websites, other recommending algorithms etc. Further steps would be adjustment of our methods to be deployable in the online environment and combine them with our previous work e.g. using linked open data [18].

Acknowledgments.   This work was supported by the grant SVV-2015-260222, P46 and GAUK-126313. The SQL export of the bookshop dataset used during the experiments can be obtained on http://www.ksi.mff.cuni.cz/~peska/bookshop2014.zip.

References

1. Baltrunas, L., Amatriain, X.: Towards time-dependant recommendation based on implicit feedback. In: CARS 2009 (RecSys)
2. Claypool, M., Le, P., Wased, M., Brown, D.: Implicit interest indicators. In: IUI 2001, pp. 33–40. ACM (2001)
3. Cremonesi, P., Garzotto, F., Turrin, R.: User-centric vs. system-centric evaluation of recommender systems. In: Kotzé, P., Marsden, G., Lindgaard, G., Wesson, J., Winckler, M. (eds.) INTERACT 2013, Part III. LNCS, vol. 8119, pp. 334–351. Springer, Heidelberg (2013)
4. Desarkar, M.S., Saxena, R., Sarkar, S.: Preference relation based matrix factorization for recommender systems. In: Masthoff, J., Mobasher, B., Desmarais, M.C., Nkambou, R. (eds.) UMAP 2012. LNCS, vol. 7379, pp. 63–75. Springer, Heidelberg (2012)
5. Eckhardt, A., Horváth, T., Vojtáš, P.: PHASES: a user profile learning approach for web search. In: WI 2007, pp. 780–783. IEEE (2007)
6. Fang, Y., Si, L.: A latent pairwise preference learning approach for recommendation from implicit feedback. In: CIKM 2012, pp. 2567–2570. ACM (2012)
7. Hidasi, B., Tikk, D.: Initializing matrix factorization methods on implicit feedback databases. J. UCS **19**, 1834–1853 (2013)
8. Hu, Y., Koren, Y., Volinsky, C.: Collaborative filtering for implicit feedback datasets. In: ICDM 2008, pp. 263–272. IEEE (2008)
9. Koren, Y., Bell, R., Volinsky, C.: Matrix factorization techniques for recommender systems. Comput. IEEE Comput. Soc. Press **42**, 30–37 (2009)
10. Lai, Y., Xu, X., Yang, Z., Liu, Z.: User interest prediction based on behaviors analysis. Int. J. Digit. Content Technol. Appl. **6**(13), 192–204 (2012)
11. Lee, D.H., Brusilovsky, P.: Reinforcing recommendation using implicit negative feedback. In: Houben, G.-J., McCalla, G., Pianesi, F., Zancanaro, M. (eds.) UMAP 2009. LNCS, vol. 5535, pp. 422–427. Springer, Heidelberg (2009)

12. Lops, P., de Gemmis, M., Semeraro, G.: Content-based recommender systems: state of the art and trends. In: Recommender Systems Handbook, Springer, Heidelberg, pp. 73–105 (2011)
13. Ostuni, V.C., Di Noia, T., Di Sciascio, E., Mirizzi, R.: Top-N recommendations from implicit feedback leveraging linked open data. In: RecSys 2013, pp. 85–92. ACM (2013)
14. Peska, L.: IPIget– The Component for Collecting Implicit User Preference Indicators. In ITAT 2014, Ustav informatiky AV CR, **2014**, 22–26. http://itat.ics.upjs.sk/workshops.pdf
15. Peska, L., Vojtas, P.: Evaluating various implicit factors in e-commerce. In: RUE (RecSys) 2014, CEUR, 910, pp. 51–55 (2012)
16. Peska, L., Vojtas, P.: Negative implicit feedback in e-commerce recommender systems. In: WIMS 2013, pp. 45:1–45:4. ACM (2013)
17. Peska, L., Vojtas, P.: Recommending for disloyal customers with low consumption rate. In: Geffert, V., Preneel, B., Rovan, B., Štuller, J., Tjoa, A.M. (eds.) SOFSEM 2014. LNCS, vol. 8327, pp. 455–465. Springer, Heidelberg (2014)
18. Peska, L., Vojtas, P.: Enhancing recommender system with linked open data. In: Larsen, H. L., Martin-Bautista, M.J., Vila, M.A., Andreasen, T., Christiansen, H. (eds.) FQAS 2013. LNCS, vol. 8132, pp. 483–494. Springer, Heidelberg (2013)
19. Raman, K., Shivaswamy, P., Joachims, T.: Online learning to diversify from implicit feedback. In: KDD 2012, pp. 705–713. ACM (2012)
20. Rendle, S., Freudenthaler, C., Gantner, Z., Schmidt-Thieme, L.: BPR: bayesian personalized ranking from implicit feedback. In: UAI 2009, pp. 452–461. AUAI Press (2009)
21. Yang, B., Lee, S., Park, S., Lee, S.: Exploiting various implicit feedback for collaborative filtering. In: WWW 2012, pp. 639–640. ACM (2012)

Product Recommendation for Small-Scale Retailers

Marius Kaminskas[1], Derek Bridge[1]([⊠]), Franclin Foping[2], and Donogh Roche[2]

[1] Insight Centre for Data Analytics, University College Cork, Cork, Ireland
{marius.kaminskas,derek.bridge}@insight-centre.org
[2] NitroSell Ltd., Cork, Ireland
{franclin.foping,donogh.roche}@nitrosell.net

Abstract. Product recommendation in e-commerce is a widely applied technique which has been shown to bring benefits in both product sales and customer satisfaction. In this work we address a particular product recommendation setting — small-scale retail websites where the small amount of returning customers makes traditional user-centric personalization techniques inapplicable. We apply an item-centric product recommendation strategy which combines two well-known methods – association rules and text-based similarity – and demonstrate the effectiveness of the approach through two evaluation studies with real customer data.

Keywords: Product recommendation · Online shopping · Association rules · Text-based similarity · Hybrid approach · User study

1 Introduction

The benefits that recommender systems (RSs) can bring to e-businesses are widely recognized. In addition to direct increase of revenue, RSs have been shown to increase customer loyalty and direct customers to new items in the product catalog [4]. Well-known examples of e-commerce recommenders, such as those used by Amazon[1], attract a large user community and typically rely on user-centric recommendation techniques that exploit the target user's shopping history [9]. However, a small-scale retail setting poses additional challenges for product recommendation. Users of small-scale e-commerce websites often do not have extensive shopping history records, many customers being one-time visitors. Consequently, traditional rating-based personalization techniques (i.e., user-based or item-based collaborative filtering) are inapplicable in such settings.

In this work we propose a flexible product recommendation solution which can be applied to various product domains and which provides meaningful recommendations without relying on user profiling. We develop our approach working with two real-world websites — a party costume and accessory store which in

---

[1] http://www.amazon.com/gp/help/customer/display.html?nodeId=16465251.

H. Stuckenschmidt and D. Jannach (Eds.): EC-Web 2015, LNBIP 239, pp. 17–29, 2015.
DOI: 10.1007/978-3-319-27729-5_2

this paper we refer to as *Retailer #1*, and a skateboarding shop which we refer to as *Retailer #2*. Both businesses are small-scale retailers, *Retailer #1*'s web site receiving a daily traffic of around 900 visits on average and *Retailer #2*'s site receiving on average 200 daily visits. For both retailers, roughly 50 % of the visitors only view one product and few are returning customers. The customer-product purchase data is therefore sparse: during the first two months of the evaluation period, out of 7800 products in *Retailer's #1* catalog, 2200 items were purchased, roughly 50 % of them only once; for *Retailer #2*, out of 1500 products, 90 were purchased, out of them 70 only once.

Since such data is not sufficient for applying *user-centric* recommendation techniques, we adopt an *item-centric* approach, by establishing a degree of *relatedness* between any two products in a retailer's product catalog. We identify two techniques for computing item relatedness – one based on textual descriptions of products, and the other based on product co-occurrence in shoppers' browsing histories. The proposed recommender is based on a combination of the two techniques. Being able to compute a relatedness score for any pair of products allows us to implement a service which provides product recommendations when a user is viewing a product web page. The viewed product acts as a 'seed' or 'query' for recommending the top-N most related products from the catalog, which can be displayed in a recommendation panel on the product page.

The contributions of this work are the following: (a) analyzing the problem of product recommendation in the particular setting of small-scale retailers; (b) suggesting a technique which is applicable to any product domain (provided that the products have text descriptions); and (c) performing a user study with real customers of two retail websites.

In the following section we describe product recommendation techniques used in e-commerce. Next, we describe the implementation of the proposed approach. Then, we describe the offline experiments conducted to validate the adopted recommendation strategy. Finally, we describe the online evaluation of the system and outline future work directions.

2   Related Work

A major challenge encountered when applying RS algorithms to real world e-commerce platforms is *data sparsity* — users view or purchase only a small fraction of the product catalog thus making traditional rating-based techniques difficult to apply. Moreover, user profiling in an e-commerce setting is challenging due to the lack of explicit ratings. Due to the above challenges, e-commerce recommendations often cannot be closely tailored to the preferences of each individual user, but need to be generated in a way that would satisfy the majority of customers [6,7]. Hence, recommendations in e-commerce are typically based on computing item-to-item similarities and using these to recommend items (products) that are similar to the ones viewed or purchased by the user [3,8]. The core step in such approaches is reliably computing item similarity, which is often alleviated by employing data mining techniques, such as product clustering and association rule mining [10].

Cho et al. [3] used the shopping behaviour of online customers as a source of item similarity information. The authors distinguished three levels of user's involvement with an item — an item view, a basket placement, and an item purchase. Product association rules were mined for each source of information separately and then combined into a single item similarity score by giving most importance to item co-occurrence among purchases and least importance to item co-occurrence among viewed items. The authors also employed a product taxonomy to address the data sparsity problem — grouping products into categories (e.g., types of apparel) before association rule mining.

Li et al. [8] proposed modeling the grocery recommendation problem as a bipartite graph with users and items as nodes, and edges representing the purchase of an item by a user. The authors computed product similarity using transition probabilities between items in the graph (passing through the user nodes). While the first order of transition probabilities only allowed establishing similarity between items that were bought together, repeating the probability propagation resulted in higher orders of similarity. This allowed establishing similarity between items that did not appear in the same baskets but were related through common neighbours, thus alleviating the data sparsity problem.

Product recommendation for small-scale retailers is even more challenging compared to the large-scale retail setting, particularly due to the small number of returning customers and limited purchase history of individual users. Chen et al. [2] addressed this problem by combining product association rules with a number of heuristics for providing recommendations when the available data is not sufficient for association rule mining. The proposed heuristics included recommending products that are most popular among users from the target user's country, or products that are most frequently purchased in the last month.

In our work, we also employ association rule (AR) mining, however we address the data sparsity problem by combining ARs with text-based item similarity. Moreover, similarly to Cho et al. [3], to cope with the limited amount of purchase data, we use product views as a source for AR mining.

3   The Approach

We observe that retail websites typically organize the product data into categories containing products that are similar in terms of their intended use, for instance, the product *reindeer costume* may belong to a category *animal costumes*. We exploit such grouping when evaluating our approach in an offline setting (see Sect. 4).

Furthermore, individual products can vary according to certain characteristics (e.g., size or colour). For instance, the product *reindeer costume* may vary by size — small or large. The item *small reindeer costume* is the actual product variant sold by the retailer. Given such an organization of products, our goal was to design a recommendation service which functions on the level of products to avoid recommending variants of the same product (e.g., recommending a small reindeer costume for users viewing a large costume of the same kind).

The proposed item-centric product recommender first computes relatedness scores for any pair of products in the retailer's catalog. Then, given a product viewed by the user, the system can obtain all scores between the viewed product and other products in the catalog, rank them according to the score, and recommend the top-N products to the user. The product relatedness scores can be pre-computed, since they do not depend on the user.

We view product relatedness as either item *similarity* or *complementarity* — two Christmas-themed costumes may be considered similar to each other, while a costume and a matching accessory are complementary. The proposed text-based relatedness computation approach mostly allows capturing product similarity relations, while the co-occurrence-based approach may capture both similarity and complementarity relations. Next we describe the two approaches for computing product relatedness scores.

3.1   Text-Based Approach

The text-based similarity computation is a technique widely used in web mining, information retrieval, and natural language processing, since it allows estimating similarity between a pair of text documents and may be used for matching a user's query to documents, for document clustering, etc.

To compute the text-based relatedness of two products, we represent each product as a document concatenating the *name*, *keywords*, and *description* of the product taken from the retailer's database.

The text documents are then preprocessed using stopword removal, stemming, and tokenization, converting the documents into a *bag of n-grams* representation. The collection of all product documents is then turned into a matrix of feature vectors with one row per document (i.e., a product) and one column per feature (i.e., a token). We use Python's scikit-learn package[2] for text preprocessing and building the document matrix.

Having built the document matrix, we can compute the similarity between any pair of vectors in the matrix (i.e., documents). We define the text-based relatedness score of two products as the cosine similarity between their vector representations:

$$\mathrm{rel}_{text}(i,j) = \frac{d_i \cdot d_j}{\|d_i\| \times \|d_j\|} \tag{1}$$

where d_i and d_j are the vectors of the documents describing products i and j.

The process of text preprocessing and creating vector representations of the documents depends on a number of settings, e.g., the minimal length of terms to be considered for tokenizing the documents, the n-gram length range, etc. The optimal configuration of these settings was determined through an offline evaluation (see Sect. 4).

---

[2] http://scikit-learn.org/stable/modules/feature_extraction.html#text-feature-extraction.

3.2   Co-occurrence-based Approach

The second technique we employ for computing relatedness scores uses association rule (AR) mining. While the general form of an AR is $(X \Rightarrow Y)$ where X and Y are sets of products and the presence of items X implies a high chance of observing items Y, we limited our analysis to rules containing one product on each side, i.e., $(i \Rightarrow j)$, where i and j are products.

Since the *purchase* transaction records of small-scale retailers typically do not provide sufficient product catalog coverage, we employ product *views* for AR mining. The underlying assumption for this method is that if two products are frequently viewed in a single user session, they are related to each other.

To get the product view data, we require a log of product pages accessed by users. We implemented and deployed user tracking functionality on the websites of the two retailers in our study. We stored the acquired data as a log of product page views attributed to permanent user session IDs. The ARs are extracted from this log using the Apriori AR mining algorithm [1].

For any pair of products for which there is a rule $(i \Rightarrow j)$, we define a relatedness score between products i and j, similar to the confidence of the corresponding rule:

$$\mathrm{rel}_{AR}(i,j) = \frac{|\{S_U : i, j \in S_U\}|}{|\{S_U : i \in S_U\}|} \tag{2}$$

where S_U is the set of all user sessions, and a user session is the set of all product page views accessed by the user.

Since the AR-based approach relies on actual product views, we cannot guarantee complete coverage of the product catalog. In other words, there will be products which do not appear in any rules and therefore do not have a set of related products. In fact, for the two retailers that were involved in the study, the catalog coverage was equal to 6 % and 10 % of the products.

Moreover, even if a product does appear in a rule, it is typically found in only a few association rules; for our two retailers, those products which appear in rules appear in only 1 to 3 association rules. Since our goal is to compute the top-N related products for any given catalog product (and for any N value), we cannot rely solely on the AR-based approach for product recommendation. However, the approach proves to be valuable when combined with the text-based approach as we show in Sect. 4.

3.3   Hybrid Approach

Unlike the AR-based approach, the text-based approach is able to compute the relatedness scores for any pair of products in the catalog (assuming they all have text descriptions). Therefore, we propose a hybrid combination of the two techniques: given a product, we compute k of the top-N related products by first applying the AR-based approach ($k \in [0, N]$), and then fill the remaining $N - k$ slots with the top-ranked products returned by the text-based approach. The precedence of AR-based approach over the text-based technique was chosen because the ARs are more accurate (see Sect. 4, Table 2) and they cover both the similarity and complementarity aspects of product relatedness.

Additionally, we have implemented a hybrid approach combining association rules with product popularity. Having computed the top-k related products with the AR-based approach, we fill the remaining $N - k$ with the most popular items (popularity estimated as the number of product views).

4   Offline Evaluation

We use offline experiments to determine the optimal configuration of the text-based approach described above and to compare the different product recommendation approaches, using data of the two retailers. Both retailers use NitroSell eCommerce — a configurable shopping platform which provides product data and purchase transaction storage facilities.

NitroSell's platform provides a basic product recommendation panel displaying up to 8 product suggestions when a user is viewing a product page. Therefore, in our experiments, we set $N = 8$ when generating the top-N recommendations.

In Nitrosell's platform at present, the recommendations for each product (which populate the recommendation panels) are primarily determined manually by the retailer combined with (very limited) information about product co-occurrences among purchased items. Our aim was to improve this legacy approach to recommendation.

4.1   Experimental Setup

Evaluating the proposed product relatedness computation requires a *ground truth* of product relatedness. In other words, to evaluate the relatedness scores that our algorithms compute, we need to know which products are actually related in reality. Since such information is not directly available in retailers' datasets, we approximated it with two sources of information — the co-purchased items and items belonging to the same product theme:

1. *Co-purchased* products are pairs of products that co-occured in user baskets when they made a purchase at the online store, and these were available to us because NitroSell's platform records them in its database.
2. *Co-themed* products are related by a theme, which is manually assigned to them by the retailer, e.g., all party costumes and accessories sold during the Christmas period might be assigned a *Christmas* theme. To perform a more detailed evaluation of the recommendation approaches, we also considered two subsets of the co-themed products as distinct ground truth sources.
3. *Substitute* products belong to the same theme and the same product category. We assume a pair of such products to be substitutes for each other, e.g., two different Christmas-themed animal costumes.
4. *Complementary* products belong to the same theme, but different product categories. We assume a pair of such products to complement each other, e.g., a Christmas animal costume and a Christmas Santa costume.

The above sources of information are not available for all products in the retailers' product catalogs. Therefore, as made explicit in Table 1, we restricted the offline experiments to the products that are covered by the ground truth information and performed the experiments for each of the four product sets independently.

For each of the four types of ground truth (*Co-purchased*, *Co-themed*, *Substitute*, and *Complementary*), we denote the set of products covered by the ground truth as P and define *recall* and *precision* metrics:

$$recall = \frac{|\{p \in P : (Rel_p \cap Top_p) \neq \emptyset\}|}{|P|} \quad prec. = \frac{\sum_{p \in P} |\{i \in Top_p : i \in Rel_p\}|}{N \cdot |P|}$$

where Rel_p is the set of products related to product p according to the ground truth, and Top_p is the set of top-N products retrieved by the product relatedness computation approach. In other words, we are measuring the ratio of products for which we can correctly recover at least one related item in the ground truth, and the average ratio of correct product recommendations in top-N.

4.2   Results

As a baseline approach for comparing against the proposed recommendation approaches, we used popularity-based product selection — for any given product, the top-8 most popular (in terms of page views) products were selected. In addition to the pure text-based approach, we used the hybrid combination of AR-based and text-based techniques, and the combination of AR-based and popularity-based methods (see Sect. 3.3).

For each product recommendation approach, we computed four recall and precision values — one for each type of ground truth described in the previous section. Table 1 shows the evaluation results for *Retailer #1*. The obtained results show all proposed approaches to outperform the popularity baseline and the hybrid combination of AR and text-based techniques to outperform other methods. (Results for *Retailer #2* were analogous and are therefore omitted).

The differences in Table 1 between the pure text-based approach and the hybrid combination of the text-based and AR-based approaches are small. This is because the AR-based approach is applicable to only 6 % of *Retailer's #1* product catalog, and so its usefulness is 'lost' in the averaging of the recall values for all products in the ground truth sets.

Table 1. Recall (and precision) values for *Retailer #1*.

Approach	Co-purchased items (5020 products)	Same theme items (4445 products)	Substitutes (4170 products)	Complementaries (3085 products)
Popularity	0.16 (0.022)	0.094 (0.023)	0.005 (0.001)	0.135 (0.032)
AR + pop	0.232 (0.045)	0.185 (0.047)	0.112 (0.025)	0.141 (0.033)
Text-based	0.645 (0.278)	0.91 (0.59)	0.83 (0.475)	**0.222 (0.053)**
AR + text	**0.653 (0.284)**	**0.912 (0.591)**	**0.839 (0.478)**	**0.222 (0.053)**

Therefore, to confirm the usefulness of the AR-based approach (hence supporting selection of the AR + text hybrid), we report the metric values for each ground truth considering only products that are covered by the ARs (Table 2).

Table 2. Recall (and precision) values for products covered by ARs (*Retailer #1*).

Approach	Co-purchased items (670 products)	Same theme items (577 products)	Substitutes (547 products)	Complementaries (367 products)
Text-based	0.578 (0.141)	0.792 (0.174)	0.713 (0.158)	0.065 (0.01)
AR-based	**0.706 (0.18)**	**0.811 (0.187)**	**0.815 (0.188)**	**0.068** (0.01)

The results show a clear advantage of the pure AR-based approach over the text-based approach. This is particularly evident for the *Co-purchased* products. We conclude that the AR-based approach can correctly identify related products for the portion of the catalog that it covers. Since these products are likely to be the most popular (most frequently viewed) in the catalog, it is essential to include the AR-based approach when generating recommendations. We therefore selected the hybrid combination of the AR-based and text-based techniques to be used in the online experiments.

5   Online Evaluation

Having identified the best method of computing the product relatedness score, we deployed the proposed product recommender on the two retailers' websites, integrating the recommendation panel into NitroSell's platform.

The online evaluation of the recommender was conducted within an A/B testing framework: website users were randomly assigned to either group A or group B. Users in group A were shown the legacy version of the recommendation panel, while users in group B were shown the panel generated using the proposed technique — a hybrid combination of AR and text-based approaches. As we discussed in Sect. 4, the legacy recommendations are primarily determined manually. Therefore, the legacy version of the panel provides a non-trivial baseline for the evaluation, as we are comparing automatically generated recommendations against manually-defined ones.

5.1   Experimental Setup

To compare the effectiveness of the product recommendations in groups A and B, we identified the users by a persistent session ID. Once randomly assigned to group A or B, the users were kept in the same group for subsequent visits to the website. The experiment data was logged by recording uniquely identifiable records — *events*. Event entries consist of a number of attributes, among others:

- *eventType* defines the type of the logged event and may have the following values: {*productview, addtobasket, removedfrombasket, ordercomplete*}. These

event types correspond to the following events, respectively: the web page for the product was viewed by the user, the product was added to the user's basket, the product was removed from the basket, and the purchase of the items in the basket was completed;
- *recommendedItems* defines the list of products that were displayed in the product recommendation panel on the product's web page (applicable to events with eventType=*productview*);
- *orderTotal* denotes the value in euros of the completed order (applicable to events with eventType=*ordercomplete*);
- *timestamp* denotes the time of the logged event.

A user *session* is defined as the set of events attributed to the same session ID value. Each session can belong to only one experiment group.

5.2 Performance Metrics

For each experiment group, we computed a number of performance metrics to compare the user behavior and the effectiveness of product recommendations in the two groups. The following metrics were used in the evaluation:

- The click-through rate for the product recommendation panel, which we define as the ratio of product page views which originated from a click on a recommended product over the total number of product page views:

$$\frac{|\,e \in E_G \;:\; \text{eventType}=productview \;\&\; \text{productId} \in R_G\,|}{|\,e \in E_G \;:\; \text{eventType}=productview\,|}$$

where E_G is the set of all events in the target experiment group ($G = \{A, B\}$) and R_G is the set of all product IDs found in the *recommendedItems* attribute values among events that occurred before *e.timestamp* in the same session.
- The average number of product page views per session:

$$\frac{|\,e \in E_G \;:\; \text{eventType}=productview\,|}{|S_G|}$$

where S_G is the set of *sessions* in group G. This metric corresponds to the average session length which is a common performance metric in e-commerce.
- The average number of completed orders per session:

$$\frac{|\,e \in E_G \;:\; \text{eventType}=ordercomplete\,|}{|S_G|}$$

which corresponds to the *conversion rate* — another common performance metric for e-commerce systems.

We note that the definition above of a *recommendation click* is not strict — it does not require the user to immediately click on a recommended product, but includes product page views of the recommended item that occur later in the session. The rationale behind this is that even if users do not directly click on the recommendation, they may be driven to search for it later. A stricter definition of the recommendation click is one where we consider only product page view events whose product ID was among the recommendations in the *previous* session event. We report results for both relaxed and strict definitions.

5.3   Results

The results that we present here come from running the online experiment between March 3rd and August 25th on *Retailer's #1* website, and between March 30th and August 25th on *Retailer's #2* website. Prior to analyzing the collected data, we filtered the log to exclude duplicate events (which may occur when refreshing a webpage) and to discard user sessions that either contain no product page views, do not begin with a product page view, or consist of one event only (this indicates customers being redirected from third party shopping platforms).

The remaining data amounts to 7850 (8158) unique user sessions in group A (B resp.) for *Retailer #1*, and 1516 (1627) user sessions in group A (B resp.) for *Retailer #2*.

We first measured the recommendation panel *click-through rate* for the two websites. For *Retailer #1*, the results show a rate of 0.05 for group A and 0.1 for group B (using the strict definition of the recommendation click), and 0.16 (0.25) for group A (B resp.) using the relaxed definition. The numbers for *Retailer #2* data are 0.07 (0.19) for the strict definition and 0.17 (0.37) for the relaxed definition in groups A (B resp.). Both retailers show consistency in the results — the users are more likely to click on a recommended product when it is generated using the proposed approach. We also observe that users are more likely to click on the recommendation panel on *Retailer's #2* website. This can be explained by the different placement of the panel on the two websites: *Retailer #1* displays the panel at the bottom of the page, therefore preventing some users from seeing the panel without scrolling, while *Retailer #2* displays recommendations on the side of the screen, making them more visible to the users.

The *average session length* for both retailers is slightly higher for group B: for *Retailer #1* the sessions had an average of 6.4 page views in group A and 6.9 in group B; for *Retailer #2* the values are 4.9 (6.8) for groups A (B resp.).

The *conversion rate* for both retailers showed no difference between the experiment groups: for *Retailer #1* both groups showed an average of 0.14 orders per session, for *Retailer #2* an average of 0.04 orders per session.

To further analyze the purchase data in the two experiment groups, we restricted our analysis to users who clicked the recommendation panel at least once during their interaction with the website. Tables 3 and 4 present the number of completed orders and total revenue (in euros) among all recorded sessions, and among sessions that contain a recommendation click (*SD* – strict definition, *RD* – relaxed definition).

Table 3. An analysis of completed orders for *Retailer #1*.

Group	Num. of sessions		Num. of orders		Total revenue	
	A	B	A	B	A	B
All sessions	7850	8158	1067	1114	38907	39720
SD sessions	1713	2910	340	536	14814	20758
RD sessions	2545	3655	606	737	24688	28016

Table 4. An analysis of completed orders for *Retailer #2*.

Group	Num. of sessions		Num. of orders		Total revenue	
	A	B	A	B	A	B
All sessions	1516	1627	62	71	4735	6258
SD sessions	329	689	19	28	1285	2832
RD sessions	458	753	27	34	2168	3181

For *Retailer #1*, the total revenue numbers are approximately equal in both groups. But, when restricting the analysis to user sessions that contain a recommendation click, the total revenue is higher for group B, due to the fact that this group contains more sessions with recommendation clicks. For *Retailer #2*, the total revenue is higher for group B — both for all the user sessions, and for sessions containing a recommendation click.

To summarize, we observe that the product recommendation panel in both websites is not frequently noticed by the users. This may be influenced by the visibility of the panel, so alternative placement strategies may be explored in the future. However, among users who click on the recommendations, the number of completed orders and total revenue are higher in group B. This leads us to believe that the proposed recommendation approach brings benefit to the retailers.

6   Conclusions and Future Work

We have proposed a recommender that is a hybrid combination of two techniques of which the AR-based approach provides higher-quality recommendations but which, due to data sparsity (i.e., few products being purchased together), cannot provide recommendations for all products in the catalog. Therefore, a second technique – the text-based approach – is a necessary complement when generating recommendations for the full product catalog.

The obtained evaluation results lead us to believe that the proposed approach results in a more attractive recommendation panel, since the users are more likely to click on it compared to the legacy version of the panel. We also conclude that recommendation placement is essential, since users are more likely to click on recommendations if they are clearly visible on the website and less likely to click on them if scrolling is required. The results also showed that among users who engage with product recommendations, the number of completed orders and total revenue are higher compared to the legacy version of the recommender. Moreover, the proposed recommendation approach does not require manual input from the retailers compared to the legacy version of the recommendation panel in both websites.

An important future work direction is investigating alternative placement of product recommendations. In addition to displaying the recommendation panel on the product description page, recommendations could be made on the checkout page. This alternative placement poses interesting research questions: Are the same techniques applicable to recommendations when browsing and purchasing products? Should we take into account the active user's basket contents when generating recommendations?

Moreover, we may investigate new hybrid solutions (e.g., combining manual recommendations with the AR-based approach). Another possibility is to exploit external sources of information, such as existing product taxonomies, to enrich the text descriptions of products and to improve the quality of the text-based relatedness computation.

We are also interested in exploiting recommendation techniques for increasing sales diversity [5], as the current data suggests a power law distribution of product popularity for both retailers.

Finally, a user trial dedicated to recommendation perception could help understanding the effectiveness of the proposed techniques. In the current experiments, the users were not aware that they were part of an experiment. Actively gathering their feedback about the product recommendations could help us obtain important insights.

Acknowledgements. This research has been conducted with the financial support of Science Foundation Ireland (SFI) under Grant Number SFI/12/RC/2289.

References

1. Agrawal, R., Srikant, R., et al.: Fast algorithms for mining association rules. In: Proceedings of the 20th International Conference on Very Large Databases (VLDB), vol. 1215, pp. 487–499 (1994)
2. Chen, J., Miller, C., Dagher, G.: Product recommendation system for small online retailers using association rules mining. In: Proceedings of the International Conference on Innovative Design and Manufacturing, pp. 71–77 (2014)
3. Cho, Y.H., Kim, J.-K., Ahn, D.H.: A personalized product recommender for web retailers. In: Baik, D.-K. (ed.) AsiaSim 2004. LNCS (LNAI), vol. 3398, pp. 296–305. Springer, Heidelberg (2005)
4. Benjamin Dias, M., Locher, D., et al.: The value of personalised recommender systems to e-business: a case study. In: Proceedings of the 2008 ACM Conference on Recommender Systems, pp. 291–294. ACM (2008)
5. Fleder, D., Hosanagar, K.: Blockbuster culture's next rise or fall: The impact of recommender systems on sales diversity. Manag. Sci. **55**(5), 697–712 (2009)
6. Giering, M.: Retail sales prediction and item recommendations using customer demographics at store level. ACM SIGKDD Explor. Newsl. **10**(2), 84–89 (2008)

7. Lee, J., Hwang, S.-W., Nie, Z., Wen, J.-R.: Navigation system for product search. In: 2010 IEEE 26th International Conference on Data Engineering (ICDE), pp. 1113–1116. IEEE (2010)
8. Li, M., Dias, B.M., et al.: Grocery shopping recommendations based on basket-sensitive random walk. In: Proceedings of the 15th International Conference on Knowledge Discovery and Data mining, pp. 1215–1224. ACM (2009)
9. Sarwar, B., Karypis, G., Konstan, J., Riedl, J.: Item-based collaborative filtering recommendation algorithms. In: Proceedings of the 10th International Conference on World Wide Web, pp. 285–295. ACM (2001)
10. Schafer, J.: The application of data-mining to recommender systems. Encycl. Data Warehouse. Min. 1, 44–48 (2009)

Using Graph Metrics for Linked Open Data Enabled Recommender Systems

Petar Ristoski[✉], Michael Schuhmacher, and Heiko Paulheim

Research Group Data and Web Science,
University of Mannheim, Mannheim, Germany
{petar.ristoski,michael,heiko}@informatik.uni-mannheim.de

Abstract. Linked Open Data has been recognized as a useful source of background knowledge for building content-based recommender systems. While many existing approaches transform that data into a propositional form, we investigate how the graph nature of Linked Open Data can be exploited when building recommender systems. In particular, we use path lengths, the K-Step Markov approach, as well as weighted NI paths to compute item relevance and perform a content-based recommendation. An evaluation on the three tasks of the 2015 LOD-RecSys challenge shows that the results are promising, and, for cross-domain recommendations, outperform collaborative filtering.

Keywords: Linked Open Data · Recommender systems · Graph metrics · Cross-domain recommendation

1 Introduction

Recommender systems are systems that provide a suggestion of items to a user, based on the user's profile and/or previous behavior. They are used, e.g., for music recommendation in streaming services, in online shopping sites, or on news portals and aggregators. The two major types of recommender systems are *collaborative filtering* and *content-based* recommender systems. The former exploit similarity among *users*, i.e., they recommend items that have been consumed and/or ranked high by users that have similar interests as the user for which the recommendation is made. The latter exploit similarities among *items*, e.g., recommending music of the same genre or news articles on the same topic. Combinations of those approaches, known as *hybrid* approaches, have also been widely studied [1].

In particular for content-based recommender systems, Linked Open Data (LOD) has been shown to be a valuable source of background knowledge. Despite data from various domains being published as LOD [23], particularly cross-domain sources such as DBpedia [13] are primarily used in recommender systems. Given that the items to be recommended are linked to a LOD dataset, information from LOD can be exploited to determine which items are considered to be similar to the ones that the user has consumed in the past. For example,

© Springer International Publishing Switzerland 2015
H. Stuckenschmidt and D. Jannach (Eds.): EC-Web 2015, LNBIP 239, pp. 30–41, 2015.
DOI: 10.1007/978-3-319-27729-5_3

DBpedia holds information about genres of books and music recordings, which can be exploited in recommendation systems [9,21]. Most often, selected data is extracted from DBpedia and transformed into a propositional form, i.e., each graph node is represented by a flat vector of binary and/or numeric features. However, DBpedia contains more information than expressed in those propositional forms. In particular, semantic *paths* between entities are a good candidate for building cross-domain recommender systems.

In this paper, we step away from extracting flat, propositional content features [22] from LOD sources, and consider the graph nature of those sources instead. Specifically, we look at paths between the items as a measure for the similarity of those items. We explore different variants of such path-based similarity measures and contrast them with standard collaborative filtering methods.

The rest of this paper is structured as follows. In Sect. 2, we explicate our overall approach. Section 3 shows how our approach is applied to the three tasks of the 2015 LOD-RecSys challenge[1], i.e., top-N recommendations, diversity, and cross-domain recommendations, and discusses the results. We conclude with a review of related work in Sect. 4 and a short summary in Sect. 5.

2 Graph Methods for Recommender Systems

In this paper, we consider three graph-based recommendation approaches. To perform the calculations, we first build an undirected weighted graph, where each item is represented as a node. For our implementation, we use the JUNG java library[2], which also offers implementations of different algorithms from graph theory. Since the domains for the three tasks differ, we use the same set of graph algorithms, but a different graph of items with different edge weights for each of the tasks. The different graph algorithms are described in this section, while the dataset-specific construction of the respective graphs is described in the corresponding parts of Sect. 3.

2.1 Shortest Path

To recommend relevant items for each user, we try to find the items in the graph that are closest to those items that were liked by the user, i.e., we assume that *proximity* in the graph is a proxy for *similarity*.

For implementing that approach, let R be the set of items a user liked. Then, for each item t in the test set (i.e., the items from which a recommendation is to be made), we compute the negated sum of shortest path lengths (given the edge weights) for all items in R to t as a ranking score:

$$I(t|R) = -\sum_{i=1}^{|R|} sp(R_i, t), \tag{1}$$

---

[1] http://sisinflab.poliba.it/events/lod-recsys-challenge-2015/.
[2] http://jung.sourceforge.net/.

where $sp(R_i, t)$ is the shortest path from R_i to t, where $R_i \in R$. Then, for each user the test items are sorted based on the relevance I in descending order, and the top-N items are recommended to the user.

2.2 K-Step Markov Approach

The PageRank algorithm is frequently used to compute importance for nodes in a graph. The PageRank score of a node can be seen as the probability of visiting that node with a random walk on the graph [16]. The K-Step Markov approach represents an algorithm variant of the PageRank algorithm with priors and computes the importance of any node in a given graph based on a given root set of nodes [28]. More precisely, the approach computes the relative probability that the system will spend time at any particular node in the graph, given that it starts in a root set of nodes R and ends after K steps.

The result is an estimate of the transient distribution of states in the Markov chain, starting from R: as K gets larger it will converge to the steady-state distribution used by PageRank, i.e. the standard version of PageRank without priors. Thus, the value of K controls the relative tradeoff between a distribution "biased" towards the root set of nodes R and the steady-state distribution which is independent of where the Markov process started. The relative importance of a node t given a root set R can be calculated using the equation:

$$I(t|R) = \left[A_{P_R} + A^2_{P_R} + \ldots + A^K_{P_R} \right] \tag{2}$$

where A is the transition probability matrix of size $n \times n$, and P_R is an $n \times 1$ vector of initial probabilities for the root set R.

In order to create recommendations, we again start with the set of items R which a user likes, and then use the K-Step Markov approach to find the top-N nodes that have the highest stationary probability. For the transition probability we use the edge weights after turning them into proper probabilities.

2.3 WeightedNIPaths

For predicting the relevance of an item node for a given user within the graph, we also make use of the WeightedNIPath algorithm [28]. Building upon the Shortest Path approach from Sect. 2.1, the idea is here to consider not only the single shortest path, but to take into account all distinct paths and add a decay factor λ to penalize longer paths. Therefore, we compute the number of distinct paths between the source nodes, i.e. all nodes/items that have been rated $r \in R$ by the user, and each other node t, i.e. all unrated and potentially to be recommended nodes. Our intuition is that items which are frequently (indirectly) connected to positively rated items, have some content-based connection and should thus get a higher recommendation score by this method.

Formally, for a set of items R liked by a user, and a candidate item t, we compute the importance score as

$$I(t|R) = \sum_{r \in R} \sum_{i=1}^{|P(r,t)|} \lambda^{-|p_i|} \tag{3}$$

where $P(r, t)$ is a set of maximum-sized node-disjoint paths from node r to node t, p_i is the ith path in $P(r, t)$, and λ is the path decay coefficient. As before, the top-N items are recommended.

With this approach, movies that e.g. share the same actors and director will be closer related than two movies that have only the director in common. While being similar to the shortest path approach described above, WeightedNIPath takes into account all paths and discounts each edge of a path by a factor, thus penalizing longer paths (we use a discount factor of 3). The rationale here is that the longer a path, the less closely related are the items this path connects. In addition, based on initial experiments, we limit the path length to 2, which is also common practice when working with DBpedia as a semantic network [25].

3 Evaluation

We evaluate the item recommendation performance of the three graph algorithms with benchmark data from the Linked Open Data-enabled Recommender Systems Challenge 2015.[3] For this challenge, three training datasets from different domains, i.e., movies, books, and music, are provided. Those datasets were generated by collecting data from Facebook profiles about personal preferences ("likes") for the items. After a process of user anonymization, the items available in the dataset have been mapped to their corresponding DBpedia URIs. An overview of the size of the datasets is given in Table 1.

For all three tasks, each approach was evaluated on an unseen gold standard using an online evaluation system provided by the challenge, and compared to standard collaborative filtering as a baseline.

Table 1. Datasets overview

Dataset	#Items	#Ratings	Task
movies	5,389	638,268	1 & 3
music	6,372	854,016	2
books	3,225	11,600	3

3.1 Task 1: Top-N Recommendations from Unary User Feedback

In this task, top-N recommendations for the movie domain are to be made. The input is unary feedback (i.e., whether a user likes an item) under open world semantics, i.e., no negative examples (dislikes) are provided. The evaluation is made based on recall, precision, and F-measure for the top 10 recommendations.

---

[3] http://sisinflab.poliba.it/events/lod-recsys-challenge-2015/.

Graph Extraction. The graph we construct consists of some direct relations of the movies, as well as their actors, genres, and characters. To generate the graph, we used the RapidMiner Linked Open Data extension [19,20]. We extracted the following relations:

- **movie:** *rdf:type*, *dcterms:subject*, *dbpedia-owl:starring*, *dbpedia-owl:director*, *dbpedia-owl:distributor*, *dbpedia-owl:producer*, *dbpedia-owl:musicComposer*, *dbpedia-owl:writer*, and *dbpprop:genre*
- **movie_actor:** *rdf:type*, *dcterms:subject*, and *is dbpedia-owl:starring of*
- **movie_genre:** *rdf:type* and *dcterms:subject*
- **movie_character:** *rdf:type*, *dcterms:subject*, *dbpedia-owl:creator*, and *dbpedia-owl:series*

Each DBpedia entity is represented as a node in the graph, where the relations between the entities are represented as undirected edges between the nodes in the graph. We use inverse document frequency (IDF) to weight the edges. For example, if an actor plays in five movies, then the IDF for each of those five relations is $log\frac{1}{5}$.[4]

In order to make the approach work also for rather weakly interlinked resources, we introduce additional edges in the graph based on the abstracts in DBpedia. To that end, we preprocess the abstract, i.e., we convert the abstract to lower case, perform tokenization, stemming, and stop words removal. Then, each token is represented as a node in the graph. Eventually, we use the resulting graph of abstract tokens and genre relations to find paths between entities from the movie domain and entities from the books domain. The relations between the entities and tokens are, as before, represented as undirected edges between the nodes in the graph and edges are also again IDF-weighted as for task 1.

In addition to the graph, we extracted the following *global* (i.e., not user-related) popularity scores:

- Number of Facebook likes[5]
- Metacritic score[6]
- Rotten Tomatoes score[7]
- DBpedia Global PageRank [26]
- Local Graph PageRank: Computed using the *JUNG* java library on the previously generated graph
- Aggregated Popularity: Using Borda's rank aggregation [4], we aggregated all those popularity scores into one.

Furthermore, following the approach in [21], we also use a stacking approach [27] for combining all the recommendations, i.e., collaborative, content-based, and global.

---

[4] To compute edge weights from IDF, we first normalize the IDF scores to [0; 1], and then assign $1 - IDF_{normalized}$ as a weight to the edges, so that edges with a larger IDF value have a lower weight.

[5] https://www.facebook.com/.

[6] http://www.metacritic.com/.

[7] http://www.rottentomatoes.com/.

Table 2. Results for top-N recommendations from unary user feedback (the best results are marked in bold).

Approach	P@10	R@10	F1@10
User-Based KNN (k=20)	0.0954	0.1382	0.1129
User-Based KNN (k=50)	0.1025	0.1485	0.1213
User-Based KNN (k=80)	**0.1032**	**0.1493**	**0.122**
Shortest Path	0.0597	0.0859	0.0704
K-Step Markov Approach (K=4)	0.0496	0.0703	0.0581
WeightedNIPaths (H=2)	0.0151	0.0217	0.0178
Shortest Path (w abstract)	0.0525	0.0746	0.0616
K-Step Markov Approach (K=4) (w abstract)	0.0562	0.0804	0.0662
WeightedNIPaths (H=2) (w abstract)	0.0216	0.0309	0.0254
Borda's rank aggregation	0.0572	0.0824	0.0676
Stacking with polynomial regression	0.0099	0.0137	0.0115

Results. The results are depicted in Table 2. It can be observed that the collaborative filtering based approaches clearly outperform the content-based ones. From the graph-based approaches, the shortest paths are the best performing approach.

As already observed in [21], the global ranking scores, aggregated with Borda's rank aggregation, are a strong competitor to content-based approaches. On the other hand, the stacking approach combining multiple recommendation approaches does not work well. This is in particular due to the fact that only positive evidence is given, which makes it hard to train a regression algorithm.

Due to its bad performance, we have not considered the stacking solution for the subsequent tasks.

3.2 Task 2: Diversity Within Recommended Item Sets

The second task is to make recommendations in the music domain. Here, the focus is on *diverse* recommendations, i.e., entities from different genres. The evaluation is made based on the average of F-measure and intra-list diversity (ILD) for the top 20 recommendations.

Graph Extraction. To generate the graph, we extracted the following relations:

- **music_artist:** *rdf:type, dcterms:subject, dbpedia-owl:genre, dbpedia-owl: associatedBand, dbpedia-owl:associatedMusicalArtist, dbpedia-owl:genre,* and *dbpedia-owl:occupation*
- **music_band:** *rdf:type, dcterms:subject, dbpedia-owl:associatedBand, dbpedia-owl:associatedMusicalArtist, dbpedia-owl:genre,* and *dbpedia-owl:bandMember*

- **music_album:** *rdf:type*, *dcterms:subject*, *dbpedia-owl:artist*, and *dbpedia-owl:genre*
- **music_composition:** *rdf:type*, *dcterms:subject*, *dbpedia-owl:musicalArtist*, *dbpedia-owl:musicalBand*, and *dbpedia-owl:genre*
- **music_genre:** *rdf:type* and *dcterms:subject*
- **abstract:** *dbpedia-owl:abstract*

As for the previous task, each DBpedia entity is represented as a node in the graph, where the relations between the entities are represented as undirected edges between the nodes in the graph. Like for the previous task, we use IDF to weight the edges.

For making the predictions, with user-based k-NN, we simply predict the top 20 items, as we already observe a high ILD with this approach. For shortest paths and the k-step Markov approach, we follow the approach in [21]: We first generate ranked lists. From those lists, we then pick the top 10 items, and then iteratively fill them up with the next 10 items in the list that do not share a genre with those that are already in the list.

Table 3. Results for diversity within recommended item sets (the best results are marked in bold).

Approach	P@20	R@20	F1@20	ILD@20
User-Based KNN (k=20)	0.09	0.2477	0.1321	0.9039
User-Based KNN (k=50)	0.0963	0.2649	0.1412	0.9028
User-Based KNN (k=80)	**0.0973**	**0.2677**	**0.1427**	0.9032
Shortest Path	0.0343	0.0948	0.0504	0.8964
K-Step Markov Approach (K=4)	0.0312	0.0863	0.0458	0.9077
WeightedNIPaths (H=2)	0.0343	0.0933	0.0502	0.8588
Shortest Path (w abstract)	0.0077	0.0203	0.0112	**0.9717**
K-Step Markov Approach (K=4) (w abstract)	0.0217	0.0585	0.0317	0.9699
WeightedNIPaths (H=2) (w abstract)	0.0358	0.0975	0.0523	0.8785
Borda's rank aggregation	0.0356	0.0977	0.0522	0.922

Results. The results for task 2 are depicted in Table 3. Again, we can see that on average, the collaborative filtering approaches produce the better results in terms of F-measure, with the ILD being comparable. It is furthermore remarkable that the ILD is that high for the collaborative filtering approaches, which were not specifically altered for producing diverse recommendations. The highest ILD score is achieved with the Shortest Path (with abstract), however at the cost of a very low recall, precision, and F1 score.

3.3   Task 3: Cross-Domain Recommendation

The third task poses a different setting, as it asks for using feedback (user ratings) from one domain, here movies, to provide recommendations for another domain, namely books. Like for the first task, recall, precision, and F-measure for the top 10 recommendations are used as evaluation metrics.

Graph Extraction. To generate the graph, we extract the same features as for the top-N recommendation tasks on movies (see Sect. 3.1), but including in addition the *dbpedia-owl:abstract* for each item. For the book domain, we extract the following relations:

- **book:** *rdf:type, dcterms:subject, dbpedia-owl:genre, dbpedia-owl:author,* and *dbpedia-owl:subsequentWork*
- **book_writer:** *rdf:type* and *dcterms:subject*
- **book_character:** *rdf:type* and *dcterms:subject*
- **book_genre:** *rdf:type* and *dcterms:subject*
- **abstract:** *dbpedia-owl:abstract*

Results. The results for task 3 are depicted in Table 4. Here, in contrast to task 1 and 2, we find that the graph-based approaches clearly outperform the collaborative filtering (CF) ones. We also observe that User-based KNN is comparably low, which leads us to the suspicion that the cross-domain nature of this task poses a serious challenge to regular CF approaches – which obviously need to operate on already observed items. In an extreme cross-domain scenario, there would be no historical user preference information available on the items to be ranked and any CF approach would fail. In contrast, the graph-based approaches can apparently find reasonable relations in the graph between books and movies (e.g., common genres, or mentioning of similar terms in the abstract) and leverage those for creating meaningful predictions.

Table 4. Results for cross-domain recommendation (the best results are marked in bold).

Approach	P@10	R@10	F1@10
User-Based KNN (k=20)	0.0162	0.026	0.0199
User-Based KNN (k=50)	0.022	0.0353	0.0271
User-Based KNN (k=80)	0.0258	0.0416	0.0318
Shortest Path	0.0326	0.0539	0.0407
K-Step Markov Approach (K=4)	0.0659	0.1077	0.0818
WeightedNIPaths (H=2)	0.0358	0.0299	0.0227
Shortest Path (w abstract)	0.0627	0.1026	0.0778
K-Step Markov Approach (K=4) (w abstract)	**0.078**	**0.1276**	**0.0968**
WeightedNIPaths (H=2) (w abstract)	0.0195	0.0314	0.024
Borda's rank aggregation	0.0301	0.0493	0.0374

4   Related Work

It has been shown that LOD can improve recommender systems towards a better understanding and representation of user preferences, item features, and contextual signs they deal with. LOD has been used in content-based, collaborative, and hybrid techniques, in various recommendation tasks, i.e., rating prediction, Top-N recommendations and diversity in content-based recommendations. An overview of cross-domain recommender systems is given in [2].

Among the earliest such efforts is *dbrec* [17], which uses DBpedia as knowledge base to build a content-based music recommender system. The recommendations are based on measure (named Linked Data Semantic Distance) which computes the distance between two items in the DBpedia graph, using only the object properties. Heitmann et al. [9] propose an open recommender system which utilizes Linked Data to mitigate the new-user, new-item and sparsity problems of collaborative recommender systems. They first publish an existing music artists database as LOD using the FOAF ontology[8]. Then, they link the data to DBpedia and DBtune MySpace. Using the new connections between the users, artists and items, the authors are able to build a collaborative recommender system.

Di Noia et al. [5] propose a model-based recommender system that relies on LOD, and can use any arbitrary classifier to perform the recommendations. First, they map the items from the local dataset to DBpedia, and then extract the direct property-value pairs. The resulting data is converted to feature vectors, where each property-value pair represents a feature. The work is extended in [6] where the similarities between the items are calculated using a vector space model. In this approach, for each item the direct property-value pairs are extracted from DBpedia, Freebase and LinkedMDB[9], which are represented as a 3-dimensional matrix where each slice refers to an ontology property and represents its adjacency matrix. In [15], the authors present SPrank, a hybrid recommendation algorithm for computing top-N item recommendations from implicit feedback exploiting the information available as LOD. In the approach, the authors try to extract features able to characterize the interactions between users, items and entities capturing the complex relationships between them. To do so, they extract all the paths that connect the user to an item in order to have a relevance score for that item, which are then used as features in the recommender algorithm. In more recent work [14], the authors propose a content-based recommender based on a neighborhood-based graph kernel, which computes semantic item similarities by matching their local neighborhood graphs.

In [21], we present a hybrid multistrategy book recommender system, where we use stacking regression and rank aggregation to combine the results of multiple base recommenders. The features for the recommenders are generated using the previously described RapidMiner LOD extension, from multiple LOD sources. Another approach by Schmachtenberg et al. [24] uses background knowledge from LinkedGeoData, to enhance a location-based recommendation system.

---

[8] http://xmlns.com/foaf/spec/.
[9] http://www.linkedmdb.org/.

The features for the recommendation system were generated using the FeGeLOD tool [18], the predecessor of the RapidMiner LOD extension.

Furthermore, several cross-domain recommender systems based on LOD data have been proposed in the literature. Fernández-Tobías et al. [7] proposed an approach that uses DBpedia as a cross-domain knowledge source for building a semantic network that links concepts from several domains. On such a semantic network, which has the form of a directed acyclic graph, a weight spreading activation algorithm [3] retrieves concepts in a target domain (music) that are highly related to other input concepts in a source domain (points of interest). The work is extended in [11,12] by finding richer semantic relations between architecture and music concepts in DBpedia.

A similar LOD-enhanced graph-based approach is presented in [8,10]. The approach is based on an enhanced spreading activation model that exploits intrinsic links between entities across a number of data sources.

5   Conclusion

In this paper, we have proposed to derive weighted graphs from DBpedia, and apply graph algorithms on it to retrieve item-based recommendations. We studied the usage of three different graph algorithms working on different subgraphs of the DBpedia graph. Our approaches rely on (a) shortest paths, (b) a variant of PageRank with priors (K-Step Markov), and (c) the sum of all distinct, connecting paths (WeightedNIPaths).

We find that in situations where the complete user feedback is available, collaborative filtering outperforms all studied graph-based approaches. In contrast, in situations where user feedback is scarce – here: for making cross-domain predictions – graph-based approaches are a reasonable way to build recommender systems. This observation makes the approaches proposed in this paper an interesting candidate for various settings. For example, in cold start situations, where no ratings for new products exist (yet), they should be included in recommendations. Second, when trying to open new market segments for an existing customer base, such methods can be helpful.

So far, we have considered only DBpedia as LOD source. In future work we can explore the existing *owl:sameAs* links in DBpedia to build richer and denser graphs from domain specific LOD sources, e.g., LinkedMDB[10] for movies, MusicBrainz[11] for music, the British National Bibliography[12] for books, etc. The information retrieved from the domain specific LOD sources should be more accurate and extensive, which should lead to better performance of the recommender systems. Furthermore, when building the graphs only specific relations were included, which we believed were the most relevant for the task. However, the graphs may be built in unsupervised manner, i.e., including all properties for all of the entities, and expanding the graph to several hops. The proposed

---

[10] http://www.linkedmdb.org/.

[11] https://wiki.musicbrainz.org/LinkedBrainz.

[12] http://bnb.data.bl.uk/.

approaches should still be able to make good recommendations on such graphs, because we use IDF to weight the edges, i.e., the most relevant edges will have a higher weight. This way, we would be able to apply the approaches on data from any domain/s without the need for manual feature engineering.

Acknowledgements. The work presented in this paper has been partly funded by the German Research Foundation (DFG) under grant number PA 2373/1-1 (Mine@LOD). Part of this work was performed on the computational resource bwUniCluster funded by the Ministry of Science, Research and the Arts Baden-Württemberg and the Universities of the State of Baden-Württemberg, Germany, within the framework program bwHPC. We would like to thank our colleague Robert Meusel for his valuable contribution to our system.

References

1. Burke, R.: Hybrid recommender systems: Survey and experiments. User Model. User-Adapted Interact. **12**(4), 331–370 (2002)
2. Cantador, I., Fernández-Tobías, I., Berkovsky, S., Cremonesi, P.: Cross-domain recommender systems (2015)
3. Allan, M.: Collins and Elizabeth F Loftus. A spreading-activation theory of semantic processing. Psychol. Rev. **82**(6), 407 (1975)
4. de Borda, J.C.: Mémoire sur les élections au scrutin. Histoire de l'Academie Royale des Sciences (1781)
5. Di Noia, T., Mirizzi, R., Ostuni, V.C., Romito, D.: Exploiting the web of data in model-based recommender systems. In: Proceedings of the Sixth ACM Conference on Recommender Systems, RecSys 2012, pp. 253–256. ACM, New York, NY, USA (2012)
6. Di Noia, T., Mirizzi, R., Ostuni, V.C., Romito, D., Zanker, M.: Linked open data to support content-based recommender systems. In: Proceedings of the 8th International Conference on Semantic Systems, I-SEMANTICS 2012, pp. 1–8. ACM, New York, NY, USA (2012)
7. Fernández-Tobías, I., Cantador, I., Kaminskas, M., Ricci, F.: A generic semantic-based framework for cross-domain recommendation. In: Proceedings of the 2Nd International Workshop on Information Heterogeneity and Fusion in Recommender Systems, HetRec 2011, pp. 25–32. ACM, New York, NY, USA (2011)
8. Heitmann, B., Dabrowski, M., Passant, A., Hayes, C., Griffin, K.: Personalisation of social web services in the enterprise using spreading activation for multi-source, cross-domain recommendations. In: AAAI Spring Symposium: Intelligent Web Services Meet Social Computing (2012)
9. Heitmann, B., Conor Hayes, C.: Using linked data to build open, collaborative recommender systems. In: AAAI Spring Symposium: Linked Data Meets Artificial Intelligence (2010)
10. Heitmann, B., Hayes, C.: SemStim at the LOD-RecSys 2014 challenge. In: Presutti, V., et al. (eds.) SemWebEval 2014. CCIS, vol. 475, pp. 170–175. Springer, Heidelberg (2014)
11. Kaminskas, M., Fernández-Tobías, I., Ricci, F., Cantador, I.: Knowledge-based identification of music suited for places of interest. Inf. Technol. Tourism **14**(1), 73–95 (2014)

12. Kaminskas, M., Fernández-Tobías, I., Cantador, I., Ricci, F.: Ontology-based iden-
    tification of music for places. In: Cantoni, L., (Phil) Xiang, Z. (eds.), Information
    and Communication Technologies in Tourism 2013, pp. 436–447. Springer, Heidel-
    berg (2013)
13. Lehmann, J., Isele, R., Jakob, M., Jentzsch, A., Kontokostas, D., Mendes, P.N.,
    Hellmann, S., Morsey, M., van Kleef, P., Auer, S., Bizer, C.: DBpedia - a large-scale,
    multilingual knowledge base extracted from wikipedia. Seman. Web J. (2013)
14. Ostuni, V.C., Di Noia, T., Mirizzi, R., Di Sciascio, E.: A linked data recommender
    system using a neighborhood-based graph kernel. In: Hepp, M., Hoffner, Y. (eds.)
    EC-Web 2014. LNBIP, vol. 188, pp. 89–100. Springer, Heidelberg (2014)
15. Ostuni, V.C., Di Noia, T., Mirizzi, R., Di Sciascio, E.: Top-n recommendations
    from implicit feedback leveraging linked open data. In: IIR, pp. 20–27 (2014)
16. Page, L., Brin, S., Motwani, R., Winograd, T.: The pagerank citation ranking:
    Bringing order to the web. Technical Report 1999–66, Stanford InfoLab, November
    1999. Previous number = SIDL-WP-1999-0120
17. Passant, A.: dbrec — Music recommendations using DBpedia. In: Patel-Schneider,
    P.F., Pan, Y., Hitzler, P., Mika, P., Zhang, L., Pan, J.Z., Horrocks, I., Glimm, B.
    (eds.) ISWC 2010, Part II. LNCS, vol. 6497, pp. 209–224. Springer, Heidelberg
    (2010)
18. Paulheim, H., Fürnkranz, J.: Unsupervised generation of data mining features from
    linked open data. In: International Conference on Web Intelligence, Mining, and
    Semantics (WIMS 2012) (2012)
19. Paulheim, H., Ristoski, P., Mitichkin, E., Bizer, C.: Data mining with background
    knowledge from the web. In: RapidMiner World (2014)
20. Ristoski, P., Bizer, C., Paulheim, H.: Mining the web of linked data with rapid-
    miner. J. Web Seman. (2015). To appear
21. Ristoski, P., Loza Mencía, E., Paulheim, H.: A hybrid multi-strategy recommender
    system using linked open data. In: Presutti, V., et al. (eds.) SemWebEval 2014.
    CCIS, vol. 475, pp. 150–156. Springer, Heidelberg (2014)
22. Ristoski, P., Paulheim, H.: A comparison of propositionalization strategies for cre-
    ating features from linked open data. In: Linked Data for Knowledge Discovery
    (2014)
23. Schmachtenberg, M., Bizer, C., Paulheim, H.: Adoption of the linked data best
    practices in different topical domains. In: Mika, P., et al. (eds.) ISWC 2014, Part
    I. LNCS, vol. 8796, pp. 245–260. Springer, Heidelberg (2014)
24. Schmachtenberg, M., Strufe, T., Paulheim, H.: Enhancing a location-based rec-
    ommendation system by enrichment with structured data from the web. In: Web
    Intelligence, Mining and Semantics (2014)
25. Schuhmacher, M., Ponzetto, S.P.: Knowledge-based graph document modeling. In:
    Proceedings of the 7th ACM International Conference on Web Search and Data
    Mining, WSDM 2014, pp. 543–552. ACM, New York, NY, USA (2014)
26. Andreas Thalhammer. Dbpedia pagerank dataset. Downloaded from (2014).
    http://people.aifb.kit.edu/ath/#DBpedia_PageRank
27. Ting, K.M., Witten, I.H.: Issues in stacked generalization. Artif. Intell. Res. **10**(1),
    271–289 (1999)
28. White, S., Smyth, P.: Algorithms for estimating relative importance in networks.
    In: Proceedings of the Ninth ACM SIGKDD International Conference on Knowl-
    edge Discovery and Data Mining (KDD), pp. 266–275. ACM, New York, NY, USA
    (2003)

Multimedia Recommendation

Toward Building a Content-Based Video Recommendation System Based on Low-Level Features

Yashar Deldjoo[✉], Mehdi Elahi, Massimo Quadrana, and Paolo Cremonesi

Politecnico di Milano, Milan, Italy
{yashar.deldjoo,mehdi.elahi,massimo.quadrana,paolo.cremonesi}@polimi.it
http://www.polimi.it

Abstract. One of the challenges in video recommendation systems is the *New Item* problem, which happens when the system is unable to recommend video items, that no information is available about them. For example, in the popular movie-sharing websites, such as *Youtube*, everyday, hundred millions of hours of videos are uploaded and big portion of these videos may not contain any meta-data, to be used by the system to generate recommendations.

In this paper, we address this problem by proposing a method, that is based on automatic analysis of the video content in order to extract a number representative low-level visual features. Such features are then used to generate personalized content-based recommendations. Our evaluation shows that our proposed method can outperform the baselines, by producing more relevant recommendations. Hence, a set low-level features extracted automatically can be more descriptive and informative of the video content than a set of high-level expert annotated features.

Keywords: Recommender systems · Content based · Low level · Video

1 Introduction

Recommender Systems (RSs) are tools and techniques that suggest to users, a set of items that may be of their interest [25]. Several approaches have been already proposed and used for recommendation generation [3,11,26,28]. *Content-based* recommendation [5,22] is the classical approach that suggests items based on their associated features. For instance, news recommender systems consider the terms in the news articles as features and recommend to user the news articles that have features similar to the ones the user preferred before.

In order to generate this type of recommendations, the system must have some information about the items, beforehand. Accordingly, the system may not be able to recommend items that are new and no information is available about them. For example, in video recommendation, it may not be feasible for the system to recommend videos that no meta-data is given. Such meta-data can be of different forms, such as, the movie genre, the cast, date of production, reviews, etc.

© Springer International Publishing Switzerland 2015
H. Stuckenschmidt and D. Jannach (Eds.): EC-Web 2015, LNBIP 239, pp. 45–56, 2015.
DOI: 10.1007/978-3-319-27729-5_4

In this paper, we propose exploitation of low-level visual features, extracted from videos, in order to generate relevant recommendations. This can be used in two scenarios: (i) *New Item* scenario, i.e., there are videos, such as user generated videos, that the system has no content rather than the video file itself, and (ii) *Existing Item* scenario, i.e., some information is available for videos such as description, genre, or cast and the low-level features are used in order to improve the quality of the recommendation system. Indeed, to the best of our knowledge, all the very few related works [30,31], focused only on the second scenario. This is while, the new item scenario, is even more important to address, since in such scenario, the typical recommender systems may completely fail to generate personalized recommendations for users.

In this paper, we mainly focus on the first scenario, i.e., new item scenario, and propose a method, that automatically extracts the low-level visual features from the video content and use it for recommendation propose. We form and test the following hypothesis: a content-based recommender system, which uses a set of representative visual features of video contents, may have led to a higher accuracy in comparison to the genre based recommender system. Our offline evaluation, described later, has shown promising results, and verified our hypothesis.

The main contributions of the paper are the followings:

- we propose a method to remedy the (extreme) *New Item* problem [14] in video recommendation domain, i.e., when a new video item is added to the database, with absolutely no meta-data provided
- we assume a more realistic scenario, i.e., an up-and-running video recommender with thousands of users rather than only tens of users, that has been typically considered in the related work
- we propose a novel application of the video classification in the recommendation systems, that has been explored marginally
- we test our proposed method with the state-of-the-art evaluation methodology and measure its performance with respect to a well known *Recall* metric

The rest of the paper is organized as follows: The next section reviews the research works that are related to content-based recommender systems and existing video recommender systems. Afterwards, in Sect. 3 we describe our novel method for representing the videos based on low-level visual features, as well as our recommendation algorithm in detail. In Sect. 4 we describe the offline evaluation strategy, we conducted, to compare our proposed method with other competing methods, and in Sect. 5 we discuss the obtained results. Finally in Sect. 6, we conclude the paper and outline the future work.

2   Related Work

2.1   Content-Based Recommender Systems

Content-based recommender systems analyze a set of descriptions of the items, previously rated by a user, to build a profile of her preferences and interests according to the attributes of the objects rated by her. Indeed, recommendations are generated by matching up the attributes of the user profile (i.e., a structured representation of her interests) against the attributes of a item. In order to do so, most of the content-based recommender systems build a Vector Space Model (VSM) representation of item features. Each item is represented by a vector in a n-dimensional space, where each dimension represents an attribute from the overall set of attributes used to describe the items. Using this model, the system computes a relevance score that represents the user's degree of interest toward that item [19]. For example, in a movie recommender system, the features that represents an items can be actors, director, or genre. This strict connection with the description of items in the catalogue, also allows content-based recommender systems to produce explanations to recommendations and to naturally handle the new item problem [14].

There are various content-based recommendation algorithms. For example, classical "k-nearest neighbor" approach (KNN) computes the interest of a user for an unseen item by comparing it against all the items seen by the user in the catalogue. Each seen item contributes to predict the interest score in a way proportional to its similarity with the unseen item; this similarity is computed by means of a similarity function like *cosine similarity* or *Pearson correlation* over items' VSM representation [7,20]. Other approaches try to model the probability for the user to be interested to a target item using a Bayesian approach [21], or exploits other techniques adapted from Information Retrieval like the Relevance Feedback method [4].

Regardless of which recommendation algorithm is used, in media recommendation, the recommender system can generate recommendation based on two different types of item attributes (or features): i.e., *High-Level* (or semantic) features (HL) or *Low-Level* features (LL). The high-level features can be collected both from structured sources, such as databases, lexicons and ontologies, and from unstructured sources, such as reviews, news articles, item descriptions and social tags [4,7,12,20,21]. The low-level features, on the other hand, can be extracted directly from the media itself. For example, in music recommendation many acoustic features, e.g. rhythm and timbre, can be extracted and used to find perceptual similar tracks [8,9,17,27].

2.2   Video Recommendation and Retrieval

In video recommendation, a few works in the past have leveraged the low-level features directly extracted from the visual content itself within the recommendation process [18,30,31]. Yang et al. [30] presented a video recommender system, VideoReach, which combines textual, visual and aural video features to increase click-through-rate. Zhao et al. [31] propose a multi-task learning algorithm to integrate multiple ranking lists generated by exploring different information sources, visual content included. However, none of these previous works has considered how visual features can effectively replace the other typical content information when they are not available. Indeed, they did not address the new item scenario, where no or very little information about a video is provided to a recommender system. Instead, they considered the scenario where the low-level content is given in addition to other information and it is used to improve the quality of the recommendation. However, in this paper we address the extreme new item problem where absolutely no information is available for a video, and the system may fail to recommend this video to the users.

It worth noting that, while usage of low-level feature based video representation has been studied marginally in recommender systems community, it has been extensively researched in the other communities such as Computer Vision [24], and it has been used in a number of similar applications such as Content-Based Video Retrieval systems (CBVR). In this case, although the objectives of content-based video retrieval and video recommendation system might be different [30], they share a main approach for dealing with their specific problems which is searching for the best informative features that can represent a video. Hence, we also review briefly the literature in related research areas.

A few comprehensive surveys can be found in [10,16]. These surveys provide a good frame of reference for reviewing the literature related to video content analysis providing a large body of low-level features that can be used for video content analysis. These features are derived from either visual, auditory or textual modalities or combination of them. For example, in [24], Rasheed et al. proposed a practical movie genre classification scheme based on solely computable visual cues. In [23], the authors proposed a similar approach by considering also the audio features. Finally, in [32] Zhou et al. propose a framework for automatic classification using a temporally-structured feature based on intermediate level representation of scenes.

3   Method Description

The first step in order to build a content-based video recommendation system is search for the features that can bridge the gap between high-level concepts and low-level contents in videos. These features must comply with human norms of perception and abide by the grammar of the film - the rules creators of movies use to make a movie. In general, a movie M can be represented by three main modalities, visual, audio and text $M = M(M_V, M_A, M_T)$. In this work, we only focus on visual features, therefore

$$M = M(M_V) \tag{1}$$

where the visual modality M_V can be represented by a set of features

$$M_V = M_V(f_v) \tag{2}$$

where $f_v = (f_{v1}, f_{v2}, ..., f_{vn})$ is a set of n features obtained from the visual content. By carefully studying the features commonly used in the literature, we selected the features studied by the authors in the vision community [24] under mild modifications. We later analyzed the accuracy of features selected by performing a classification analysis and features selection based on exhaustive search.

3.1 Visual Features

A total of four main low-level visual features were used in our experiment from which a feature vector of length six ($n = 6$) was extracted to represent each video. They include

- *Average shot length*: A shot is a single camera action and the number of shots in a video can provide useful information about the pace at which a movie is being created. Average shot length is defined by

$$\overline{L}_{sh} = \frac{n_f}{n_{sh}} \tag{3}$$

where n_f is the number of frames and n_{sh} the number of shots in a movie. For example, action movies usually contain rapid movements of the camera (therefore they contain higher number of shots or shorter shot lengths) compared to dramas which often contain conversations between people (thus longer average shot length). Because movies can be made a different frame rates, $\overline{L}_{sh}$ is further normalized by the frame rate of the movie.
- *Color variance*: The variance of color has a strong correlation with the genre. For instance, directors tend to use a large variety of bright colors for comedies and darker hues for horror films. For each key frame represented in Luv color space we compute the covariance matrix:

$$\rho = \begin{pmatrix} \sigma_L^2 & \sigma_{Lu}^2 & \sigma_{Lv}^2 \\ \sigma_{Lu}^2 & \sigma_u^2 & \sigma_{uv}^2 \\ \sigma_{Lv}^2 & \sigma_{uv}^2 & \sigma_v^2 \end{pmatrix} \tag{4}$$

The generalized variance can be used as the representative of the color variance in each key frame given by

$$\Sigma = det(\rho) \tag{5}$$

in which a key frame is a representative frame within a shot (e.g. the middle shot).
- *Motion*: Motion within a video can be caused mainly by the camera movement (*i.e.* camera motion) or movements on part of the object being filmed (*i.e.* object motion). While the average shot length captures the former characteristic of a movie, it is desired for the motion feature to also capture the latter characteristic. A motion feature descriptor based on optical flow [6, 15] was used which provides a robust estimate of the motion in sequence of images based on velocities of images being filmed. Because motion features are based upon sequence of images, they are calculated over the entire frames rather on solely key frames.
- *Lightening*: Lightening is another distinguishing factor between movie genres in such a way that the director use it as a factor to control the type of emotion they want to be induced to a user. For example, comedy movies often adopt lightening which has abundance of light (*i.e.* high gray-scale mean) with less contrast between the brightest and dimmest light (*i.e.* high gray-scale standard deviation). This trend is often known as *high-key* lightening.

On the other hand, horror movies or noir films often pick gray-scale distributions which is low in both gray-scale mean and gray-scale standard deviation, known by *low-key* lightening. In order to capture both of these parameters, after transforming all key-frames to HSV color-space [29], we compute the mean μ and standard deviation σ of the value component which corresponds to the brightness. The scene lightening key ξ defined by Eq. 6 is used to measure the lightening of key frames

$$\xi = \mu.\sigma \tag{6}$$

For instance, comedies often contain key-frames which have a well distributed gray-scale distribution which results in both the mean and standard deviation of gray-scale values to be high therefore for comedy genres one can state $\xi > \tau_c$, whereas for horror movies the lightening with poorly distributed lighting the situation is reverse and we will have $\xi < \tau_h$. In the situation where $\tau_h < \xi < \tau_c$ other movie genres (*e.g.* Drama) exists where it is hard to use the above distinguish factor for them.

3.2   Recommendation Algorithm

To generate recommendations using our Low-Level descriptors we adopted a classical "k-nearest neighbor" content-based algorithm. Given a set of users U and a catalogue of items I, a set of preference scores r_{ui} has been collected. Moreover, each item $i \in I$ is associated to its feature vector $\boldsymbol{f_i}$. For each couple of items i and j, a similarity score s_{ij} is computed using *shrunk cosine similarity* as follows

$$s_{ij} = \frac{\boldsymbol{f_i}^T \boldsymbol{f_j}}{\|\boldsymbol{f_i}\| \|\boldsymbol{f_j}\| + \lambda} \tag{7}$$

where $\lambda > 0$ is the shrinkage factor. For each item i the set of its nearest neighbors NN_i is built, $|NN_i| < K$. Then, for each user $u \in U$, the predicted preference score $\hat{r_{ui}}$ for an unseen item i is computed as follows

$$\hat{r_{ui}} = \frac{\sum_{j \in NN_i, r_{uj} > 0} r_{uj} s_{ij}}{\sum_{j \in NN_i, r_{uj} > 0} s_{ij}} \tag{8}$$

4   Evaluation Methodology

We have formulated the following hypothesis: the content-based recommender system, that exploits a set of representative visual features of video contents, may have led to a higher accuracy in comparison to the genre based recommender system. Hence, we speculate that a set low-level features extracted automatically may be more informative of the video content than a set of high-level expert annotated features.

In order to test our hypothesis, we evaluate the recommendation quality of each of the considered content-based recommender system it terms of *Recall(K)*, where K is the size of the recommendation list. If a user u has N_u relevant items, the recall in its recommendation list of size K is computed as

$$Recall(K) = \sum_{u \in U} \sum_{i=1}^{k} \frac{rel(i)}{N_u} \tag{9}$$

where $rel(i) = I[r_{ui} >= 4]$ is the item relevance function and I is the indicator function. We evaluated the $Recall(K)$ Leave-One-Out Cross-Validation (LOOCV) [13]. At each step, one relevant sample (i.e., an item having rating greater than 4) is removed from each user profile. We also removed the ratings for top-10 most popular items to discount the effect of the well known popularity effect. Then, the recommendation model is built on the remaining samples and the quality of the recommendation lists is evaluated. Results are finally averaged over all the splits that have been generated.

We have used a set of movie trailers, that the were sampled randomly from all the genres, i.e., Action, Comedy, Drama and Horror. The movie titles were selected from Movielens dataset [1], and the files were obtained from *YouTube* [2]. The dataset contained over all 210 movies, 120 of which belonging to a single genre and 90 movies belonging to multiple genres.

5   Results

Figure 1 illustrates the system's recall for different content-based recommendation methods, i.e., visual feature based (Low Level-LL), genre-based (High Level-HL Genre), and Hybrid (LL-HL Genre). We performed a feature selection based on exhaustive search and chose the best configuration as shown in the figure. Comparing the results, it is clear that our proposed method, i.e., visual feature based outperforms the other methods in terms of recall. As it can be seen in Fig. 1, the recall values for all the methods, initially begins with 0 for $N = 1$, and, as N is incremented, it increases monotonically and reaches 0.36 for visual feature based (LL), 0.14 for genre based (HL genre), and 0.11 for hybrid method (LL + HL Genre), respectively. We have also performed t-test and realized that the recall values of our method (LL) is significantly higher than (HL - Genre) with p-value = 0.01912, and hybrid (LL + HL Genre) with p-value = 0.00736. However, no significant difference has been observed between the visual feature based (HL - Genre) and hybrid (LL + HL Genre) methods (p-value = 0.30581). Hence, it is clear that our proposed method can perform the best among all the other methods. In fact, it shows that, our extracted features can represent very well the videos, allowing the system to make personalized recommendations, that better match the users tastes, specially in the extreme new item cold start situation.

In addition to the figure, in Table 1, we report more detailed results, for the different system parameters, i.e., the size of the neighborhood set (k) and the size of the recommendation list (N). In this table, we report the performance when all the proposed LL features are used (not only the best feature combination).

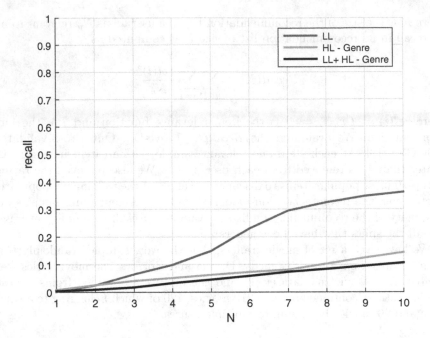

Fig. 1. Performance comparison of different CB methods under best feature combination

We have compared the results of the methods for all possible neighborhood size (k), and realized that for instance, when $N = 10$, there is an inverse proportional relation between k and recall values of LL method whereas this relation is directly proportional for hybrid method. The same condition almost exists when $N = 5$.

Moreover, it could observed that with small k, the performance of the LL method is much better than HL genre and the hybrid method for both $N = 5$ and $N = 10$. When k is high, the recall values of the hybrid method are greater than the recall values of LL and HL genre methods. However, the obtained recall values of hybrid method are still lower than the recall values of the LL method when k is small. Also, since the number of items (videos) used in our catalog is limited to only 210 items, high values of k are not justifiable for use in the knn algorithm. For these reasons, we can conclude the best performance is obtained for LL method. Indeed addition of the visual features to the genre information, do not improve the quality of the genre based method. This can be due to the fact that there is a strong correlation between the genre of a movie and the visual features that represent that movie. In fact, as noted before, it has been shown that different genres of movies, differ significantly in terms of visual characteristics.

In order to better understand this, we have also analyzed the observations and tried to classify the videos into different genres, exploiting the extracted visual features. We note that here we assumed every video to belong to only a single genre and hence considered only a subset of 120 single genre videos. We have tried different classifiers and realized that the best classification accuracy has been achieved by *Decision Tables*. We conducted 10 fold cross-validation and obtained accuracy of 73.33 %. Indeed, using the visual features, the classifier managed to successfully classify most of the videos in their correct genre. We have observed the best classification was done for comedy movies. Indeed, 27 out of 30 movies were successfully classified in their corresponding comedy genre. On the other hand, the most erroneous classification happened for the horror genre. Indeed, 8 out of 30 horror movies have been mistakenly classified as action genre. This is a phenomenon, that was expected, since typically there are many action scenes occurred in horror movies, and this may make the classification very hard.

Table 1. Performance comparison of different CB methods, in terms of Recall metric, for different neighborhood size (k) and recommendation list size (N) when all LL features are used.

	k = 4			k = 5			k = 10			k = 15		
	LL	HL	hybrid	LL	HL	hybrid	LL	HL	hybrid	LL	HL	hybrid
N = 5	**0.1476**	0.0669	0.0400	**0.1385**	0.0552	0.0393	0.0806	0.0575	**0.0835**	0.0731	0.0581	**0.1239**
N = 10	**0.2286**	0.1113	0.0628	**0.2044**	0.1025	0.0831	**0.1448**	0.1339	0.1224	0.1198	0.1430	**0.1891**

Having considered all the results, we remark that our considered hypothesis has been successfully validated, i.e., a proper extraction of the visual features of videos may have led to higher accuracy of video recommendation, than the typical expert annotation method. Indeed, it is very promising to achieve higher accuracy with automatic method than a manual method (i.e., experts analyses and annotation of videos) since the later method is very costly and in some cases even impossible (e.g., in huge datasets).

Finally, it is worth noting that, our results has been obtained by using the trailer of the movies that are only a small sample of videos themselves. Watching the trailers of even few movies, one can simply notice that the structure of them are not far different and hence the trailers of the movies do actually share many similarities which makes it much more difficult for our method to work properly. Indeed, it is much more difficult to use low-level visual features to classify movies or generate relevant movie recommendations, with their trailers, than the movies themselves. Hence, achieving high accuracy in either of the tasks, indicates the great effectiveness of our proposed method.

6   Conclusion and Future Work

In this paper, we have address the *New Item* problem by presenting a novel content-based method for video recommendation task. The proposed method extracts and uses the low-level visual features from video content in order to provide a user with personalized recommendations, without relying on any high-level features, such as, meta-data, genre, cast, or reviews, that are more costly to collect and are not available in new item cold-start situation.

We have developed the a research hypotheses, i.e., a proper extraction of the visual features of videos may have led to higher accuracy of video recommendation, than the typical expert annotation method. Based on the experiments, we conducted, we successfully verified the hypothesis and shown that the recommendation accuracy is significantly higher when using the low-level visual features than high-level genre data.

Our future work comprises the further analysis with bigger and different datasets, that we will prepare, in order to better understand the performance differences among the compared methods. We would like to also investigate the impact of using different content-based recommendation algorithms, such as those based on Bayesian, or SVD, on the performance of our method. We would like to also include additional sources of information, such as, audio features, in order to farther improve the quality of our content based recommendation method. Last but not least, we plan to perform a feature selection study in order to better understand the role and importance of the features in the performance of the CB video recommendation algorithm(s).

Acknowledgments. This work is supported by Telecom Italia S.p.A., Open Innovation Department, Joint Open Lab S-Cube, Milan.

References

1. Datasets – grouplens. http://grouplens.org/datasets/, Accessed: 01 May, 2015
2. Youtube. http://www.youtube.com. Accessed: 01 April, 2015
3. Adomavicius, G., Tuzhilin, A.: Toward the next generation of recommender systems: a survey of the state-of-the-art and possible extensions. IEEE Trans. Knowl. Data Eng. **17**(6), 734–749 (2005)
4. Ahn, J.-W., Brusilovsky, P., Grady, J., He, D., Syn, S.Y.: Open user profiles for adaptive news systems: help or harm? In: Proceedings of the 16th international conference on World Wide Web, pp. 11–20. ACM (2007)
5. Balabanović, M., Shoham, Y.: Fab: content-based, collaborative recommendation. Commun. ACM **40**(3), 66–72 (1997)
6. Barron, J.L., Fleet, D.J., Beauchemin, S.S.: Performance of optical flow techniques. Int. J. Comput. Vis. **12**(1), 43–77 (1994)
7. Billsus, D., Pazzani, M.J.: User modeling for adaptive news access. User Model. User-Adap. Inter. **10**(2–3), 147–180 (2000)
8. Bogdanov, D., Herrera, P.: How much metadata do we need in music recommendation? a subjective evaluation using preference sets. In: ISMIR, pp. 97–102 (2011)
9. Bogdanov, D., Serrà, J., Wack, N., Herrera, P., Serra, X.: Unifying low-level and high-level music similarity measures. IEEE Trans. Multimedia **13**(4), 687–701 (2011)
10. Brezeale, D., Cook, D.J.: Automatic video classification: a survey of the literature. IEEE Trans. Syst. Man Cybern. Part C Appl. Rev. **38**(3), 416–430 (2008)

11. Burke, R.: Hybrid recommender systems: Survey and experiments. User Model. User-Adap. Inter. **12**(4), 331–370 (2002)
12. Cantador, I., Szomszor, M., Alani, H., Fernández, M., Castells, P.: Enriching ontological user profiles with tagging history for multi-domain recommendations (2008)
13. Deshpande, M., Karypis, G.: Item-based top-n recommendation algorithms. ACM Trans. Inf. Syst. (TOIS) **22**(1), 143–177 (2004)
14. Elahi, M., Ricci, F., Rubens, N.: Active learning strategies for rating elicitation in collaborative filtering: a system-wide perspective. ACM Trans. Intell. Syst. Technol. (TIST) **5**(1), 13 (2013)
15. Horn, B.K., Schunck, B.G.: Determining optical flow. In: 1981 Technical Symposium East, pp. 319–331. International Society for Optics and Photonics (1981)
16. Hu, W., Xie, N., Li, L., Zeng, X., Maybank, S.: A survey on visual content-based video indexing and retrieval. IEEE Trans. Syst. Man Cybern. Part C Appl. Rev. **41**(6), 797–819 (2011)
17. Knees, P., Pohle, T., Schedl, M., Widmer, G.: A music search engine built upon audio-based and web-based similarity measures. In: Proceedings of the 30th annual international ACM SIGIR conference on Research and Development in Information Retrieval, pp. 447–454. ACM (2007)
18. Lehinevych, T., Kokkinis-Ntrenis, N., Siantikos, G., Dogruöz, A.S., Giannakopoulos, T., Konstantopoulos, S.: Discovering similarities for content-based recommendation and browsing in multimedia collections
19. Lops, P., De Gemmis, M., Semeraro, G.: Content-based recommender systems: state of the art and trends. In: Ricci, F., Rokach, L., Shapira, B., Kantor, P.B. (eds.) Recommender Systems Handbook, pp. 73–105. Springer, Heidelberg (2011)
20. Middleton, S.E., Shadbolt, N.R., De Roure, D.C.: Ontological user profiling in recommender systems. ACM Trans. Inf. Syst. (TOIS) **22**(1), 54–88 (2004)
21. Mooney, R.J., Roy, L.: Content-based book recommending using learning for text categorization. In: Proceedings of the Fifth ACM Conference on Digital libraries, pp. 195–204. ACM (2000)
22. Pazzani, M.J., Billsus, D.: Content-based Recommendation Systems. In: Brusilovsky, P., Kobsa, A., Nejdl, W. (eds.) Adaptive Web 2007. LNCS, vol. 4321, pp. 325–341. Springer, Heidelberg (2007)
23. Rasheed, Z., Shah, M.: Video categorization using semantics and semiotics. In: Rosenfeld, A., Doermann, D., DeMenthon, D. (eds.) Video Mining, pp. 185–217. Springer, Heidelberg (2003)
24. Rasheed, Z., Sheikh, Y., Shah, M.: On the use of computable features for film classification. IEEE Trans. Circ. Syst. Video Technol. **15**(1), 52–64 (2005)
25. Ricci, F., Rokach, L., Shapira, B.: Introduction to recommender systems handbook. In: Ricci, F., Rokach, L., Shapira, B., Kantor, P.B. (eds.) Recommender Systems Handbook, pp. 1–35. Springer Verlag, Heidelberg (2011)
26. Ricci, F., Rokach, L., Shapira, B.: Introduction to recommender systems handbook. In: Ricci, F., Rokach, L., Shapira, B., Kantor, P. (eds.) Recommender Systems Handbook, pp. 1–35. Springer Verlag, Heidelberg (2011)
27. Seyerlehner, K., Schedl, M., Pohle, T., Knees, P.: Using block-level features for genre classification, tag classification and music similarity estimation. Submission to Audio Music Similarity and Retrieval Task of MIREX 2010 (2010)
28. Su, X., Khoshgoftaar, T.M.: A survey of collaborative filtering techniques. Adv. Artif. Intell. **2009**, 4:2 (2009)
29. Radu, V.: Application. In: Radu, V. (ed.) Stochastic Modeling of Thermal Fatigue Crack Growth. ACM, vol. 1, pp. 63–70. Springer, Heidelberg (2015)

30. Yang, B., Mei, T., Hua, X.-S., Yang, L., Yang, S.-Q., Li, M.: Online video recommendation based on multimodal fusion and relevance feedback. In: Proceedings of the 6th ACM International Conference on Image and Video Retrieval, pp. 73–80. ACM (2007)
31. Zhao, X., Li, G., Wang, M., Yuan, J., Zha, Z.-J., Li, Z., Chua, T.-S.: Integrating rich information for video recommendation with multi-task rank aggregation. In: Proceedings of the 19th ACM International Conference on Multimedia, pp. 1521–1524. ACM (2011)
32. Zhou, H., Hermans, T., Karandikar, A.V., Rehg, J.M.: Movie genre classification via scene categorization. In: Proceedings of the International Conference on Multimedia, pp. 747–750. ACM (2010)

Personalized and Context-Aware TV Program Recommendations Based on Implicit Feedback

Paolo Cremonesi, Primo Modica, Roberto Pagano, Emanuele Rabosio[(✉)], and Letizia Tanca

Dipartimento di Elettronica, Informazione e Bioingegneria, Politecnico di Milano, Piazza Leonardo Da Vinci 32, 20133 Milano, Italy
{paolo.cremonesi,roberto.pagano,emanuele.rabosio, letizia.tanca}@polimi.it, modica1@hotmail.it

Abstract. The current explosion of the number of available channels is making the choice of the program to watch an experience more and more difficult for TV viewers. Such a huge amount obliges the users to spend a lot of time in consulting TV guides and reading synopsis, with a heavy risk of even missing what really would have interested them. In this paper we confront this problem by developing a recommender system for TV programs. Recommender systems have been widely studied in the video-on-demand field, but the TV domain poses its own challenges which make the traditional video-on-demand techniques not suitable. In more detail, we propose recommendation algorithms relying exclusively on implicit feedback and leveraging context information. An extensive evaluation on a real TV dataset proves the effectiveness of our approach, and in particular the importance of the context in providing TV program recommendations.

1 Introduction

Television is one of the most popular media in our era, and with the advent of digital TV and the growing offer of satellite services there are at any time of the day hundreds of available TV programs to be watched by the users on hundreds of different channels. On the one hand the user is satisfied by this abundance since the vast choice of programs supports his/her tastes, but on the other hand he/she suffers an information overload problem. This information overload makes the user prone to a tedious channel surfing in order to find what he/she really likes, inevitably leading to annoyance.

In the past the solution was represented by paper channel guides that used to be consulted on a daily basis. Nowadays, these paper supports are fallen into disuse due to the proliferation of channels and shows and the advent of smart

This research is partially supported by the IT2Rail project funded by European Union's Horizon 2020 research and innovation program under grant agreement No: 636078, and by the Italian project SHELL CTN01_00128_111357 of the program "Cluster Tecnologici Nazionali".

© Springer International Publishing Switzerland 2015
H. Stuckenschmidt and D. Jannach (Eds.): EC-Web 2015, LNBIP 239, pp. 57–68, 2015.
DOI: 10.1007/978-3-319-27729-5_5

TVs and smart devices, and the show schedule information has been embedded into the television software itself through the so-called *electronic program guide (EPG)*. However, the low quality of the EPG in terms of content and its often crude user interface brings to a poor user experience and, as a natural consequence, to ineffectiveness. The answer to this problem consists in providing the user with a short list of recommended programs, representing the subset of the on-air ones that most correspond to his/her preferences.

Recommendation of TV programs is rather a special instance of recommendation for three reasons:

- *Available items change over time*: many TV programs, e.g. the movies, are often broadcast once and then not anymore for a long time. The system must be able to provide recommendations also for items of this kind, if they meet the users' interests.
- *Time-constrained catalog of items*: differently from the more usual video-on-demand setting, programs are transmitted in a predefined schedule. Therefore, the recommendations must consider only the items on air at the moment in which they are requested.
- *The user feedback is usually implicit*, provided in the form of watched/not watched shows.

Note that the first issue makes it impossible to adopt traditional collaborative filtering (CF) recommendation techniques. Indeed, they are not able to recommend new items since such items cannot be compared with the other ones in terms of the feedback provided by the users in the past [1].

Moreover, a fundamental aspect to be considered in TV program recommendation is the *context* [2], i.e. the situation that the user is experiencing when watching television. The context may be characterized by a number of dimensions, the most common being the time. Other contextual information often available is represented by the social setting in which the user is accessing the content, and his/her current interest topic. So, for instance, when alone during daytime the user might prefer different shows with respect to those liked when with friends in the evening.

In this paper we propose a context-aware TV recommender system relying exclusively on implicit feedback. To the best of our knowledge, this is the first attempt to tackle both context-awareness and implicit feedback in the TV domain. The proposed techniques have been extensively evaluated on a real dataset related to Italian television.

The paper is structured as follows. Section 2 surveys the existing literature, while Sect. 3 introduces the framework in which our algorithms are supposed to operate. Section 4 describes the proposed recommendation techniques, and Sect. 5 their experimental evaluation. Finally, Sect. 6 concludes the paper.

2   Related Work

Recommendation of TV programs has raised some interest in the recent literature. The existing proposals can be divided on the basis of their aim:

recommending to build a personalized video recorder (PVR), or recommending to build a personalized EPG in linear television.

A personalized video recorder is a system generating recommendations about TV content that will be stored into an internal hard disk, for a possible future viewing by the user. The work of Engelbert et al. [3] characterizes TV programs with attributes extracted from an EPG, containing information about channel, title, subtitle, genre, actors, year and description. Recommendations of programs to be recorded are generated on the basis of an initial user profile and an adaptive user profile, both sets of TV programs classified as liked or disliked. The initial user profile is manually filled by the users, while the adaptive one is built using implicit and explicit feedback collected after the user has watched the programs. Once defined the users' profiles, the attributes of new programs (taken from the EPG) are compared against those of the programs in the user's profile with the help of a bayesian classifier. Another personalized video recorder is defined by Kurapati et al. [4]; they too propose algorithms for PVRs coupling explicit and implicit feedback, in this case relying on neural networks to combine them. The problem analyzed in these works is related to ours but is not the same, because in PVRs the recommendations do not have to be provided at specific time instants.

In the scope of linear TV, our scenario of interest, Chang et al. [5] provide guidelines to create a TV program recommender, identifying the main needed modules and performance requirements. However, the proposed framework is interesting, but just sketched; among the full-fledged proposals, just a few rely on contextual information.

Some non-contextual linear TV recommenders have appeared in the literature, and many of them rely on hybrid (collaborative and content-based) systems. Barragans-Martinez et al. [6] exploit a hybrid approach to solve new-item, cold-start, sparsity and overspecialization problems; their method uses both implicit and explicit feedback, and mixes together the outcome of content-based filtering, computed using the cosine similarity between item feature vectors, and collaborative filtering, exploiting singular value decomposition. Ali et al. [7] develop TiVo, a television-viewing service for the US market incorporating a recommender system which exploits an item-item form of collaborative filtering mixed with bayesian content-based filtering; the system envisages client and server components, and relies on both implicit feedback and explicit ratings. Another hybrid approach is that of Cotter et al. [8], who present a personalized EPG; users manually input their preferences about channels and genres, and this information is combined with the user's viewing activity by means of case-based reasoning and collaborative filtering techniques. Uberall et al. [9], on the contrary, propose a fully content-based technique, exploiting both the viewing behavior and explicit user preferences on preferred genres, subgenres and TV programs.

Context is taken into account by Ardissono et al. [10], who develop a content-based system able to generate a personalized program guide. In order to model the user, the system employs several information sources: users' explicit preferences, estimates on viewing preferences using program categories and channels, viewing preferences of stereotypical viewers classes, socio-demographic information, and users' viewing behavior. Different modules of the system manage the different kinds of information, and the results are then combined; the context is considered by the module that estimates user preferences on the basis of the user viewing behavior, since those preferences depend on day and time. Another contextual system is that of Hsu et al. [11]. They propose a hybrid system that combines the collaborative and content-based components by means of a neural network; the contextual information employed by the system is the user's mood, considered as a strong influencing factor in program selection.

All the described approaches to linear TV recommendations, both the contextual and the non-contextual ones, exploit some form of explicit feedback which must be provided by the users, like, for instance, user ratings. On the contrary, the system we propose relies only on the availability of implicit feedback in terms of history of the past program views, which is the most realistic situation. Moreover, our algorithms exploit context information in a different and more general way with respect to what is done in [10] and [11]. In fact, [10] and [11] deal only with specific kinds of context information, while we devise a framework that can accommodate every type of context dimensions, like the kind of people present during the program view or the fact that it is a weekday or the weekend.

3   System Architecture

The architecture we propose for our recommender system is shown in Fig. 1. The user interacts with a smart TV, and is allowed to request the generation of recommendations; recommendations can be generated also when the TV is turned on. The request is forwarded to the recommendation engine, that exploits the log of the past user's syntonizations, along with the context of the user and the EPG, in order to determine the list of the top-N programs to be recommended among those currently on air. Note that some kinds of contextual information may be automatically determined by the system, like the time, but others might need to be manually declared by the user, like the people with whom he/she is watching TV or his/her current mood.

4   Recommendation Methodology

Let us consider a set $\mathcal{U}$ of users and a set $\mathcal{I}$ of items, i.e. TV programs. Each item is described by n attribute dimensions $\mathcal{A}_1, \ldots, \mathcal{A}_n$, like its genre or the channel on which it is broadcast; we denote by $\mathcal{A}_j(i)$ the value of the attribute $\mathcal{A}_j$ for the item i. The context is described by m context dimensions $\mathcal{C}_1, \ldots, \mathcal{C}_m$, like time or mood.

Given the log of the past program views, we build offline, as a model to generate the recommendations, an $(n+m+1)$-dimensional tensor T, storing the number of seconds spent by the users watching the TV programs with all the possible attribute values in all the possible contexts. In more detail, consider a user u, attribute values $\mathcal{A}_1 = a_1, \ldots, \mathcal{A}_n = a_n$, and a context represented by the dimension values $\mathcal{C}_1 = c_1, \ldots, \mathcal{C}_m = c_m$. The value $t_{ua_1 \ldots a_n c_1 \ldots c_m}$ stored in the tensor represents the number of seconds the user u has spent watching programs described by attribute values $\mathcal{A}_1 = a_1, \ldots, \mathcal{A}_n = a_n$ when in context $\mathcal{C}_1 = c_1, \ldots, \mathcal{C}_m = c_m$.

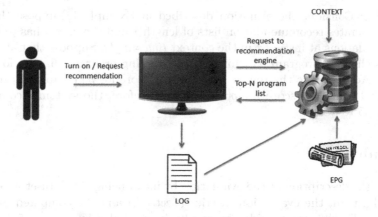

Fig. 1. System architecture

Example 1. Consider a set of context dimensions including only the time, and a set of TV program attributes constituted only by the channel. The possible values for the time context dimension are daytime and night, while those for the channel are Ch-1 and Ch-2. Figure 2 shows a possible log of syntonizations for user u.

program_id	time	channel	seconds
p1	daytime	Ch-1	2000
p2	night	Ch-2	1000
p3	daytime	Ch-1	3000
p4	night	Ch-1	2000

$$\begin{array}{c} \quad daytime \quad night \\ \begin{array}{c} Ch\text{-}1 \\ Ch\text{-}2 \end{array} \left(\begin{array}{cc} 5000 & 2000 \\ 0 & 1000 \end{array} \right) \end{array}$$

Fig. 2. Log of example 1

Fig. 3. Projection of the tensor of Example 1 for user u

In this example the tensor T has three dimensions: user, time and channel. Figure 3 shows the projection of the tensor on time and channel for user u.

Once the tensor model above has been built, it can be used at runtime to generate the recommendations. The user u requests recommendations in a given time instant t when in context $\mathcal{C}_1 = c_1, \ldots, \mathcal{C}_m = c_m$. Let $\mathcal{I}_t$ be the set of programs on air at time instant t. The system extracts from the tensor the appropriate score $r_{uic_1 \ldots c_m}$ for each item $i \in \mathcal{I}_t$, as follows:

$$r_{uic_1 \ldots c_m} = t_{u\mathcal{A}_1(i) \ldots \mathcal{A}_n(i)c_1 \ldots c_m} \tag{1}$$

If N recommendations are required, the system retrieves the N programs with the highest values of $r_{uic_1 \ldots c_m}$.

Example 2. Consider the situation described in Example 1, suppose that the system generates recommendation lists of length 1 and the user u has requested recommendations at instant t in the context *time=night*. Suppose that at instant t Ch-1 is showing program $p5$ while Ch-2 is showing program $p6$, therefore $\mathcal{I}_t = \{p5, p6\}$. According to the tensor in Fig. 1, the score for $p5$ computed using Eq. (1) is 2000 while the score for $p6$ is 1000. Therefore, the system recommends program $p5$ to u.

5 Evaluation

We start the description of the evaluation by introducing the dataset we employ (Sect. 5.1), then, the evaluation metrics (Sect. 5.2) and the compared methods (Sect. 5.3). Finally, we provide the results in a tabular form (Sect. 5.4), along with a detailed analysis (Sects. 5.5 and 5.6).

5.1 Dataset

We employed a dataset containing TV viewing information related to 7921 users and 119 channels, broadcast both over the air and by satellite. The dataset is composed by an EPG containing the description of 21194 distinct programs, and a log of the program views performed by the users. The attributes available for each program in the EPG are its genre and the channel on which it is transmitted.

The log of program views spans from December 3rd, 2013 to March 1st, 2014, and contains 10313499 entries. We deemed the syntonizations shorter than three minutes as not relevant, retaining 6525541 log entries. Each log row specifies the identifier of the user and that of the program he/she watched, along with the start time, the end time and the people with whom the user watched the program. The latter three pieces of information were used to determine the values of the three context dimensions that we chose: day of the week, time slot and *familiar context*, where with familiar context we mean just the people with whom the user was watching TV. More precisely, start and end time were employed to derive the day of the week and the time slot, where the available values for the time slot are shown in Table 1. We identified five possible relevant values for the familiar context, summarized in Table 2, depending on the age of the people; persons older than 15 years were considered adults.

The log was split in a training set, including the syntonizations between March 3rd, 2013 and February 15th, 2014 (5438977 entries), and a test set, containing the remaining ones (1086564 entries). The former was used to build the model, while the latter to assess the quality of the recommendations.

5.2 Evaluation Metrics

The performance of our recommendation algorithm was evaluated using *Recall@N*, describing the number of test items which have been included in a

Table 1. Time slots

Start time	End time	Description
02:00:00	07:00:00	Graveyard slot
07:00:00	09:00:00	Early morning
09:00:00	12:00:00	Morning
12:00:00	15:00:00	Daytime
15:00:00	18:00:00	Early fringe
18:00:00	20:30:00	Prime access
20:30:00	22:30:00	Prime time
22:30:00	02:00:00	Late fringe

Table 2. Familiar contexts

Familiar context
Mixed: adults + children
Group of adults
Group of children
Adult alone
Child alone

recommendation list of length N computed in the instant in which the viewing of the test items started and in the context in which they have been watched.

More formally, let v be a program view in the test set, v_t the start time of the view, v_u the user that watched the program, v_i the program watched and v_c the context in which the view took place. $TopN(u, c, t)$ is the set of top-N items for the user u in context c among those on air at time instant t, determined with the recommendation methodology to be evaluated. Recall@N is computed as follows:

$$Recall@N = \frac{|v \in \text{Test Set} : v_i \in TopN(v_u, v_c, v_t)|}{|v \in \text{Test Set}|} \tag{2}$$

We executed experiments for N=1, N=3 and N=5.

5.3 Compared Methods

We executed our algorithm, from now on dubbed *CtxOrd*, using different combinations of context dimensions and program attributes, with the aim of evaluating their usefulness in the generation of the recommendations. In particular, we tested the non-contextual alternatives which build the tensor T exploiting the sole channel and the sole genre. Then, we tried to enrich the tensor with the various context dimensions.

Our algorithm was compared also with a naive non-contextual and non-personalized methodology, dubbed *TopPop*, recommending to each user, in each context, the list of programs broadcast on the N channels that were globally the most seen.

Finally, we considered another less trivial, non-contextual and non-personalized competitor, named *ShortestTimeSinceStart*, always compiling the recommendation list with the programs started since the shortest time.

We had performed some trials also using traditional collaborative filtering. However, as explained in the introduction of the paper, the dynamism of the item catalog makes such techniques ill-suited for TV program recommendation, and indeed the obtained results were extremely poor. Therefore, we do not show collaborative filtering results in the following sections.

Table 3. Recall@1

Algorithm	All	≥ 8 chan	≥ 28 chan
TopPop	19.26 %	17.65	10.28 %
ShortestTimeSinceStart	3.46 %	15.89 %	7.34 %
CtxOrd – Channel	33.04 %	22.93 %	10.92 %
CtxOrd – Genre	13.91 %	12.87 %	5.85 %
CtxOrd – Channel, Genre	31.57 %	13.96 %	6.56 %
CtxOrd – Fam. Cont., Channel	33.95 %	31.92 %	19.33 %
CtxOrd – Fam. Cont., Genre	8.56 %	7.94 %	3.22 %
CtxOrd – Fam. Cont., Channel, Genre	33.20 %	22.94 %	13.96 %
CtxOrd – Day, Time, Channel	*39.23 %*	*37.02 %*	*22.36 %*
CtxOrd – Day, Time, Genre	18.11 %	16.75 %	1.44 %
CtxOrd – Day, Time, Channel, Genre	32.83 %	30.84 %	19.16 %
CtxOrd – Fam. Cont., Day, Time, Channel	39.11 %	36.34 %	21.77 %
CtxOrd – Fam. Cont., Day, Time, Genre	17.38 %	15.96 %	7.58 %
CtxOrd – Fam. Cont., Day, Time, Channel, Genre	33.29 %	31.31 %	18.88 %

All the experiments were repeated three times, considering three different compositions of the test set:

- The whole test set (7921 users, 1086564 program views).
- Subset obtained excluding the users who have shown to be not very active, having watched only 7 channels or less (5824 users, 959141 program views).
- Subset obtained including only the very active users, having watched 28 channels or more (201 users, 51525 program views).

5.4 Results

In this subsection we present in a tabular form the results obtained with the experimented methodologies. Tables 3, 4 and 5 show, respectively, Recall@1, Recall@3 and Recall@5. The tables are divided in two parts: the upper one shows the non-contextual techniques while the lower one shows the contextual ones.

5.5 Result Analysis

In the following the results reported in the tables are analyzed in detail, starting with the whole test set and then considering the reduced ones.

Full Test Set. A first aspect which can be noticed from the results is that the differences between context-aware and baseline methodologies are larger when recommending few items. This happens because many users watch just a limited

Table 4. Recall@3

Algorithm	All	≥ 8 chan	≥ 28 chan
TopPop	45.19 %	43.17 %	24.98 %
ShortestTimeSinceStart	11.33 %	33.03 %	17.77 %
CtxOrd – Channel	63.12 %	41.22 %	23.38 %
CtxOrd – Genre	30.83 %	28.78 %	13.93 %
CtxOrd – Channel, Genre	60.29 %	35.51 %	19.06 %
CtxOrd – Fam. Cont., Channel	63.94 %	61.38 %	38.93 %
CtxOrd – Fam. Cont., Genre	20.72 %	19.41 %	8.53 %
CtxOrd – Fam. Cont., Channel, Genre	49.47 %	46.44 %	29.86 %
CtxOrd – Day, Time, Channel	*67.05 %*	*64.59 %*	*42.11 %*
CtxOrd – Day, Time, Genre	37.94 %	35.72 %	7.33 %
CtxOrd – Day, Time, Channel, Genre	60.98 %	58.20 %	37.88 %
CtxOrd – Fam. Cont., Day, Time, Channel	66.94 %	62.12 %	40.70 %
CtxOrd – Fam. Cont., Day, Time, Genre	36.66 %	34.42 %	17.47 %
CtxOrd – Fam. Cont., Day, Time, Channel, Genre	62.00 %	59.26 %	38.14 %

Table 5. Recall@5

Algorithm	All	≥ 8 chan	≥ 28 chan
TopPop	59.93 %	57.81 %	31.14 %
ShortestTimeSinceStart	18.07 %	45.27 %	25.51 %
CtxOrd – Channel	77.38 %	51.19 %	30.64 %
CtxOrd – Genre	42.07 %	39.17 %	19.83 %
CtxOrd – Channel, Genre	65.18 %	48.53 %	27.72 %
CtxOrd – Fam. Cont., Channel	77.93 %	74.09 %	51.23 %
CtxOrd – Fam. Cont., Genre	73.89 %	30.14 %	14.09 %
CtxOrd – Fam. Cont., Channel, Genre	63.17 %	60.26 %	39.81 %
CtxOrd – Day, Time, Channel	78.58 %	*76.56 %*	*52.65 %*
CtxOrd – Day, Time, Genre	49.82 %	47.14 %	15.45 %
CtxOrd – Day, Time, Channel, Genre	75.31 %	72.96 %	48.90 %
CtxOrd – Fam. Cont., Day, Time, Channel	*78.67 %*	72.30 %	51.00 %
CtxOrd – Fam. Cont., Day, Time, Genre	48.26 %	45.59 %	24.32 %
CtxOrd – Fam. Cont., Day, Time, Channel, Genre	76.29 %	74.03 %	49.67 %

number of channels, and therefore even simple strategies are able to identify the proper program in lists containing several items.

Let us consider the non-contextual alternatives, above the horizontal line in the tables. We immediately note that the non-personalized methods TopPop and ShortestTimeSinceStart show very poor performance, while the personalized model based on the channel obtains very high recall. This suggests that the users' preferences are more important than the time elapsed since the program started to determine the right suggestion. Note also that the results for the personalized model relying on the genre are not good. The usage of the genre seems to confuse the system instead of helping; in fact, adding the genre to the model based on the channel brings disturbance instead of improvement.

Consider now the contextual models, below the horizontal line in the tables. First, we observe that also in this case the models with the channel behave better than those envisaging the genre, that again seems to confuse the system. The addition of the familiar context brings some improvements, but these are really small: the recall increase is less than 1 % with respect to the model based only on the channel. A significant gain, on the contrary, is provided by the usage of date and time. The best-performing model – the one including day, time and channel – improves the non-contextual alternative based on the channel of 6.19 % for Recall@1, 3.93 % for Recall@3 and 1.20 % for Recall@5; as explained above, the shorter the recommendation list, the larger the recall increment. The addition of the familiar context to the model envisaging day, time and channel does not provide significant improvements, with the exception of Recall@5.

The fact that the best model is the one envisaging day, time and channel, together with the good performance shown by the non-contextual model with only the channel, suggests that the habit factor is very important in the choices of TV viewers: many users watch very often the same channels in the same time slots.

Test Sets Obtained Excluding the Less Active Users. In this case we note that the differences between the methodologies are wide also for Recall@3 and Recall@5: this happens because the users in these test sets are used to see many channels, and so it may be difficult to discover the right program to be suggested even through long lists of recommendations.

Moreover, the negative results of the models including the genre of the program are confirmed also in this case.

In general, for each experimented model, the recall value decreases with respect to that measured with the same models on the whole test set, again because these users have seen several channels and so the recommendation is more difficult. An exception is represented by the non-personalized methodology ShortestTimeSinceStart, for which the recall obtained for the active users is greater than that achieved on the full test set. This is an interesting result, and seems to suggest that the active users are more resolved in the choice of TV programs: they know what they want to watch and change the channel when they know it is starting. The other users, on the contrary, seem to proceed in a more random way among the few channels they are used to take into account.

The two subsets of active users confirm that the contextual strategies show better performance than the non-contextual ones. The best model is again that envisaging day, time and channel, and the increment with respect to the rec-ommendations generated considering only the channel is even larger than that registered with the full test set. For instance, in the test set containing only the users having seen at least 28 channels, the increments are 11.44 % for Recall@1, 18.73 % for Recall@3 and 22.01 % for Recall@5.

Differently from what we observed in the experiments with the full test set, in this case the familiar context introduces a significant increment in the quality of recommendations. For instance, when the test set containing only the users having seen at least 28 channels is taken into account, the increments with respect to the model envisaging only the channel are 8.41 % for Recall@1, 15.55 % for Recall@3 and 20.54 % for Recall@5. However, the contribution of the familiar context is canceled when the familiar context is considered in addition to day and time. This means that the effect of the users' habits remains stronger than the impact of the familiar context, also for the active users.

5.6 Summary of the Evaluation

The described experiments showed that our methodology can provide accurate recommendations to TV users relying exclusively on implicit feedback. In addition, the experiments proved that in the considered scenario the context is decisive in the recommendation process. In more detail, the day and time context dimensions showed to be relevant for all the users, while the familiar context proved significant only for the most active users.

6 Conclusion

This paper has proposed a content-based context-aware technique to provide TV program recommendations relying exclusively on implicit feedback. An extensive evaluation on a real TV dataset has been carried out, showing the effectiveness of the proposal.

Several directions for future works exist. First of all, in this work we have exploited counters of the number of seconds which the users have spent in seeing certain programs in certain conditions, weighting each second in the same way. However, after some time it is possible that the TV is let turned on when the user has started other activities or has fallen asleep. Therefore, it could be interesting to modify the construction of our model by introducing a decay factor able to weight more the first seconds of view. Other relevant research possibilities concern the study of strategies to increase novelty and serendipity of TV recommendations.

References

1. Burke, R.: Hybrid web recommender systems. In: Brusilovsky, P., Kobsa, A., Nejdl, W. (eds.) Adaptive Web 2007. LNCS, vol. 4321, pp. 377–408. Springer, Heidelberg (2007)
2. Dey, A.K.: Understanding and using context. Pers. Ubiquit. Comput. 5(1), 4–7 (2001)

3. Engelbert, B., Blanken, M.B., Kruthoff-Brüwer, R., Morisse, K.: A user supporting personal video recorder by implementing a generic bayesian classifier based recommendation system. In: Proceedings of the PerCom Workshops, pp. 567–571. IEEE Computer Society, Los Alamitos (2011)

4. Kurapati, K., Gutta, S., Schaffer, D., Martino, J., Zimmerman, J.: A multi-agent tv recommender. In: Proceedings of the 1st International Workshop on Personalization in Future TV (2001)

5. Chang, N., Irvan, M., Terano, T.: A TV program recommender framework. In: Proceedings of KES 2013, 17th International Conference in Knowledge Based and Intelligent Information and Engineering Systems, pp. 561–570. Elsevier, Amsterdam (2013)

6. Barragáns-Martínez, A.B., Pazos-Arias, J.J., Fernández-Vilas, A., García-Duque, J., López-Nores, M.: What's on TV tonight? An efficient and effective personalized recommender system of TV programs. IEEE Trans. Consum. Electron. **55**(1), 286–294 (2009)

7. Ali, K., van Stam, W.: TiVo: making show recommendations using a distributed collaborative filtering architecture. In: Proceedings of KDD 2004, 10th International Conference on Knowledge Discovery and Data Mining, pp. 394–401. ACM, New York (2004)

8. Cotter, P., Smyth, B.: PTV: Intelligent personalised TV guides. In: Proceedings of the 17th National Conference on Artificial Intelligence and 12th Conference on Innovative Applications of Artificial Intelligence, pp. 957–964. AAAI Press / The MIT Press, Cambridge (2000)

9. Überall, C., Rajarajan, M., Rakocevic, V., Jäger, R., Köhnen, C.: Recommendation index for DVB content using service information. In: Proceedings of ICME 2009, International Conference on Multimedia and Expo, pp. 1178–1181. IEEE Computer Society, Los Alamitos (2009)

10. Ardissono, L., Gena, C., Torasso, P., Bellifemine, F., Difino, A., Negro, B.: User modeling and recommendation techniques for personalized electronic program guides. In: Ardissono, L., Kobsa, A., Mark, T. (eds.) Personalized Digital Telev., pp. 3–26. Springer, Heidelberg (2004)

11. Hsu, S.H., Wen, M.-H., Lin, H.-C., Lee, C.-C., Lee, C.-H.: AIMED- A personalized TV recommendation system. In: Cesar, P., Chorianopoulos, K., Jensen, J.F. (eds.) EuroITV 2007. LNCS, vol. 4471, pp. 166–174. Springer, Heidelberg (2007)

An LDA Topic Model Adaptation
for Context-Based Image Retrieval

Hatem Aouadi[✉], Mouna Torjmen Khemakhem, and Maher Ben Jemaa

ReDCAD Laboratory, National School of Engineers of Sfax, University of Sfax,
BP 1173, 3038 Sfax, Tunisia
{hatem.aouadi,mouna.torjmen}@redcad.org, maher.benjemaa@enis.rnu.tn
http://www.redcad.org

Abstract. In the context-based image retrieval, the textual information surrounding the image plays a central role for ranking returned
results. Although this technique outperforms content-based approaches,
it may fail when the query keywords does not match the textual content of many documents containing relevant images. In addition, users
are usually not experts and provide ambiguous queries that lead to heterogeneous results. To solve these problems, researchers are trying to
re-rank primary results using other techniques such as query expansion,
concept-based retrieval, etc. In this paper, we propose to use LDA topic
model to re-rank results and therefore improve retrieval precision. We
apply this model in two levels: global level represented by the whole document containing the image and local level represented by the paragraph
containing an image (considered as a specific textual information for the
image). Results show a significant improvement over the standard text
retrieval approach by re-ranking with the LDA model applied to the local
level.

Keywords: Image retrieval · Topic model · Re-ranking · LDA

1 Introduction

Content-based image retrieval CBIR approaches have several drawbacks mainly
related to: (i) the difficulty of knowing the semantics of the image (semantic gap),
and (ii) time and memory complexity to extract and process low-level features,
select of appropriate features, etc. Therefore, new approaches are emerged among
the best known are: keyword-based image retrieval which uses textual information surrounding the image and meta-data associated to it; and multi-modal
image retrieval which combines textual and visual futures. We are interested
in this paper to the first approach where the textual retrieval methods testi-
fied great success. These methods can be applied in the documents containing
images for image retrieval task. However, the textual information surrounding
the image may not contain the query keywords making difficult to return the
corresponding images in search results by a standard retrieval model.

© Springer International Publishing Switzerland 2015
H. Stuckenschmidt and D. Jannach (Eds.): EC-Web 2015, LNBIP 239, pp. 69–80, 2015.
DOI: 10.1007/978-3-319-27729-5_6

To address the problem of query terms and document textual content mismatching, researchers attempt to reformulate the query by adding terms similar to the original query terms. The authors in [7,18] use the pseudo-relevance feedback technique (PRF) which consists of performing a textual retrieval in a first step, and eventually select the most frequent terms in the top results to perform a new search. Although this technique has the strength of being automatic and does not require external resources, its performance is strongly related to the quality of initial results. Other researchers are trying to exploit some semantic resources for the query expansion. In [9], the author uses WordNet[1] for the expansion. The others of [16] proposed some similarity measures based on WordNet and Wikipedia to extend queries with the similar concepts. WordNet is also used in [26] in conjunction with ConceptNet[2] to add synonyms from Wordnet and the semantic meaning of terms using ConceptNet. It is clear that these methods require external semantic resources which can be very helpful but are domain dependant and not available in many languages.

In this paper, we propose to extract the hidden semantic from the collection of documents using latent modelling techniques such as Latent Dirichlet Allocation Model LDA [4]. This model has the advantage of being independent of the domain and the used language. It showed its effectiveness in various tasks such as document classification and can be used in retrieval and re-ranking search results [27,29,30].

Another major limitation of the text-based image retrieval approach appears when a document matching the query keywords may contain relevant images but also other irrelevant images. All these images will receive the same relevance score from the considered relevant document and will be classified together in the top results. This common phenomenon can easily decrease the retrieval precision. As a solution to this problem, we propose to apply LDA to the most specific information level for each image. For this, we consider the paragraph containing an image as the specific information for this image in the document and applying LDA to paragraphs instead of documents. In this way, the images in the same document receive their own relevance scores.

This paper is organized as follows. In Sect. 2, we present the LDA topic model and its application in information retrieval. Section 3 describes our approach to adapt LDA in image retrieval. We discuss in Sect. 4 the obtained results and we finish in Sect. 5 with some conclusions and perspectives.

2   State of the Art

As mentioned above, we are interested in applying LDA topic model in text based image retrieval. We present in this section LDA topic model and its application in information retrieval.

---

[1] http://wordnet.princeton.edu/
[2] http://conceptnet5.media.mit.edu/

2.1 LDA Topic Model

LDA [4] is a generative probabilistic model that was primarily designed for document collections [1], and has been widely used in many other areas such as video retrieval [14], visual concept detection [23], images modelling [17], etc. This model allows to discover the set of topics in a collection of documents. Its main idea is that the documents are represented as a random mixture of several latent topics, where each topic is characterized by a distribution of words.

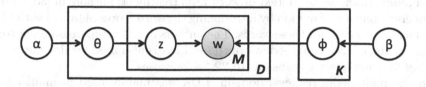

Fig. 1. Graphical representation of LDA topic model

LDA topic model can be represented with a graphical probability model as shown in Fig. 1. Suppose we have D Documents using a vocabulary of V terms. Each document is composed of M terms. Suppose also that this set of documents contains K topics. LDA provides a generative model in which each document d_i in the collection samples the topics from a K-dimensional multinomial distribution $Mul(\theta)$ and each topic Z_i samples the vocabulary terms from a multinomial distribution $Mult(\varphi)$. The estimation of φ and θ provides information about topics in the collection of documents and the weight of these topics in each document. Several algorithms have been used to estimate these parameters including variational methods [4], the Gibbs sampling model [8] and Collapsed variational inference [24].

The running of the full generative model of a simplified version of LDA algorithm is as follows:

(1) For each topic
    (a) draw a distribution over words $\varphi \sim Dir(\beta)$
(2) For each document
    (a) Chose $\theta_d \sim Dir(\alpha)$
    (b) For each word
        (i) generate topic $z \sim Mult(\theta)$
        (ii) generate term $w \sim Mult(\varphi)$.

where α and β are the Dirichlet parameters of the LDA topic model.

The model distilled textual document collections in a distribution of words that tend to co-occur in similar documents. These sets of related words are considered as "topics".

After analyzing the documents, LDA generates two probability distributions: a topic-document distribution representing the topics proportions in each document, and word-topic distribution representing the weight of words in each topic.

This reduced representation of the documents in the form of topics is very suitable for large-scale retrieval such as the Web.

2.2 LDA in Information Retrieval

Among the first studies using LDA in information retrieval, the authors in [27] applied a linear combination of word scores obtained with the language model and LDA model. This technique outperforms the use of the standard language model alone. In the same context of word level treatment, authors in [30] modify the lucene[3] similarity function by introducing the word scores obtained with LDA model as boost factor for these words. The authors in [29] compare different topic models in the information retrieval purpose. LDA is also used for documents re-ranking [31] and query expansion [10,22,28] purposes.

In the multimedia retrieval domain, LDA was mainly used in multi-modal approaches using joint distributions of visual features and words. The author in [3] describes several multimodal extensions of the basic LDA model and evaluates these models in image retrieval task. Multimodal LDA is also used in [25] as a method to identify the similarity between multimedia objects with multiple heterogeneous data sources. The authors in [13] discuss different LDA-based estimations of similarity in relation to the content based image retrieval problem.

Another application of LDA in image annotation domain [2,3,20,21] can help in image retrieval when the textual information is missed.

3   A Novel Approach of Applying LDA in Image Retrieval

In this section, we elaborate the proposed LDA based image retrieval approach. We apply this approach using the textual information in the collection of documents containing images.

Our use of LDA model in image retrieval progresses in three phases. Figure 2 illustrates the different phases of our image retrieval model.

1. **Learning phase.** This phase consists of extracting topic-distributions from the collection of documents containing images;
2. **Inference phase.** In this phase, queries should be inferred in the model learned in the first phase to estimate their topic-distributions;
3. **Retrieval phase.** We compute in this final phase the pair similarity between the query topic-distribution in one hand and the document topic-distributions on the other hand to generate results.

We detailed the first and second phases in Sect. 3.1, while the chosen similarity measure is discussed in Sect. 3.2.

---

[3] http://lucene.apache.org/core/

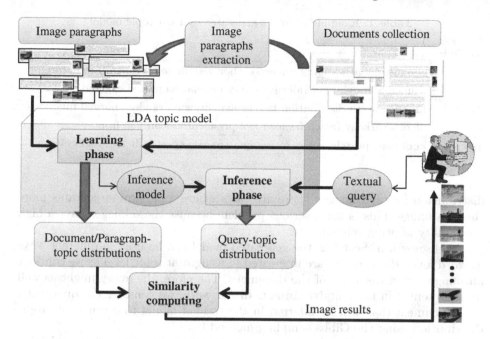

Fig. 2. LDA Image retrieval model

3.1   Adaptation of LDA Model by Taking into Account Image Specificity

Our intuition is that the image describes the global subject of a document and not the detailed information. For this reason, we use LDA topic model since it provides an overall view of the document as a distribution of topics and filters out all unnecessary information. Table 1 presents some examples of topics generated from a Wikipedia document collection using LDA topic model. Each line in the table is a topic composed of words semantically related with different degrees of relatedness. Sometimes, we can find some words non related to the topic. For example, the three last words "icon pen front" in topic t_4 are not related to the first words so not related to the topic. Thus, we can classify the topics into good and bad topics. The number of good topics reflect the quality of the used model. More we have good topics, more the used model is well.

The model generated form the document collection can be directly used to perform the retrieval process. In this case, all images in the same document will be similarly handled regardless their relevance. Furthermore, a document may discuss different topics so that images in such document are probably not related. These issues lead us to think about selecting the specific image description from the document. In this paper we consider that the paragraph containing an image is the most specific information for it. We denote this part of the document by the "local" image description and the entire document by the "global" image description. Applying LDA to the local level allows us to generate a topic

Table 1. Example of topics generated by LDA topic model

Topic id	Top terms
t_1	climate global warming science carbon greenhouse temperature nature
t_2	council borough road railway county population united england
t_3	function math distribution frac probability sum random normal functions
t_4	nicolas sarkozy french france royal presidential icon pen front
t_5	audi bmw porsche mans racing quattro cars sports tdi acura volkswagen

distribution for each image. There are cases where a paragraph contains more than an image. This is not a serious problem because these images have a high probability of being related.

As discussed in Sect. 2.2, the generated model can be used in different ways at word level. However, we are interested in using it at topic level as we consider an image reflect the topic of the document. Therefore, the query might as well be represented in the suitable format. In our work, we consider the query as a new document that will be inferred in the trained model to estimate its topic distribution using the Gibbs sampling method [8].

The final step consists in comparing the estimated query topic-distribution with the trained document (or paragraph) topic-distributions to generate results. In this paper, we chose the cosine similarity measure discussed in Sect. 3.2.

3.2 Query/Document Similarity Computing

After performing LDA on the collection of documents, it is possible to use the generated distributions to compare documents together or to infer a new document in order to estimate its distribution throughout topics. In the first case, the distributions can be used either to calculate the similarity or to calculate the divergence between documents.

There are several ways to define and calculate similarity/divergence in the LDA topic model; either between topics or between documents. In this paper, we focused on the latter. Most literature methods are based only on the document-topic distributions. For example, the work proposed in [11] measure the Kullback-Leibler divergence (KLD) [15] between these distributions, while authors in [5] introduce another dissimilarity measure based on the topic proportions. In the case of similarity computing, the authors in [13] mention the use of inner products and cosine similarity as candidates.

In scenarios where two documents have to be compared and one of them is quite shorter than the other (e.g., query and web page), the normalization of the cosine similarity shows some advantages. In our work, we are interested in determining the similarity between queries and documents. For that, we choose the cosine similarity measure.

The cosine measure is widely studied in the text retrieval. It involves measuring the cosine of the angle formed by the two vectors of the query keywords and the document terms. In our case, we have the topic proportion vectors for queries and documents. Let P and Q the probability distributions throughout topics of a document p and a query q respectively. The cosine similarity can be calculated as follows:

$$Sim_{cos}(p, q) = \frac{P.Q}{\|P\|\|Q\|} \tag{1}$$

In global search, the similarity is computed between document topic distributions and query topic distributions. The images in the most similar documents are returned as a search results. Whereas, for the local search, the similarity is computed between queries and paragraphs topic distributions to return images in the most similar paragraphs.

4   Experiments

We conducted extensive experiments on a multimedia dataset (documents containing images) to evaluate the effectiveness of LDA topic model for the image retrieval. We also compared our technique with a traditional text-based approach, query expansion techniques and a concept-based retrieval method.

4.1   Evaluation Protocols

We use in our experiments the Wikipedia collection provided by the Cross Language Evaluation Forum (CLEF) in 2011 for the *"Wikipedia Retrieval"* task. This collection consists of 125 827 documents in three different languages, containing 237 434 images with associated metadata. A set of 50 queries (textual and visual) with different difficulty levels is also provided. In this paper, we focused only on English documents whose number is 42 774 Wiki documents.

To ensure a good performance of LDA algorithm, we performed a cleaning step for the collection of documents by removing the Wikipedia style elements, infobox, tables, etc. We also reduce the number of used terms by removing stopwords, terms of overall frequency < 20 and terms of length < 3. After this cleaning phase, we apply the LDA topic model using the mallet[4]library. LDA parameters are set to the most common values used in the literature: $\alpha = 50/K$ where K is the number of topics and $\beta = 0.01$. The choice of K plays a crucial role in the effectiveness of LDA model. The authors in [6] make a first attempt to analyze this number. However, the authors assume that the data are exactly generated according to the model, and that their method may fail in the case of real world dataset. We performed several experiments with different values of K ranging from 100 to 2000 topics extracted from the entire collection. After a statistical study of its influence on the retrieval accuracy, we fixed the value of K to 1500 topics which gives better results. For text-based retrieval, we use the lucene search engine as a baseline.

To evaluate the effectiveness of our method, we are interested in the MAP (Mean Average Precision) measure which gives an idea on the overall system

---

[4] http://mallet.cs.umass.edu/

performance. It assumes that the user is interested in finding many relevant documents for each query. We also use the AP (Average Precision) measures: AP@5, AP@10 and AP@20 which evaluate results of re-ranking approaches. These measures assume that the user is interested in top results only.

4.2   Results of the Adapted Approach

We compared in a first step the text-based image retrieval with the use of LDA model alone applied to the global level (LDA document) and the local level (LDA paragraph). Table 2 shows the results.

Table 2. Evaluation results of retrieval methods

Method Name	Method Label	P@5	P@10	MAP
B	Lucene text (baseline)	**0,384**	**0,316**	**0,239**
LDA-D	LDA document	0,156	0,174	0,138
LDA-P	LDA paragraph	0,236	0,206	0,156

By comparing text-based and topic-based approaches, lucene outperforms LDA in all measures. This is not surprising due to the losses in precision form word level representation to topic level representation. In the topic level retrieval model, results tend to be more general and consequently more numerous. This can significantly reduce the precision of the retrieval model. In addition, the authors in [12] show that when the amount of information is large enough, the topic modelling becomes less efficient compared to single TF-IDF scores. In particular, a better model can be driven by the aggregation of short messages. Moreover, images in LDA paragraph model are independent from the overall document relevance.

These experiments show also that the local LDA model performs better than the global one. This may be explained by the fact that the topics trained in local level are more specific to the image then global topics of the document.

4.3   Combining Different Approaches

To use the advantages of each model, we propose to combine their results using the following formula:

$$S = \alpha \times S(M_1) + (1 - \alpha) \times S(M_2) \tag{2}$$

where $S(M_1)$ is the similarity score generated by one of the models, and $S(M_2)$ is the similarity score generated by the other. The variable α is a combination factor between 0 and 1. We tested different values of α when combining the different methods. Table 3 presents the best results based on retrieval accuracy.

Table 3. Results of Combining different retrieval methods

Combination	Best α Value	P@5	P@10	MAP
LDA-D + LDA-P	0,4	0,228	0,198	0,165 (+19% ~ to LDA)
B + LDA-D	0,8	0,380	0,298	0,244 (+2% ~ to text)
B + LDA-P	0,7	**0,400**	**0,336**	**0,281** (+17% ~ to text)
(B + LDA-D) + LDA-P	0,8	0,384	0,330	0,275 (+14% ~ to text)

The combined use of the baseline and the LDA paragraph model showed significant improvements in results of both LDA collection (+19% ($p = 0,0051$)) and tex-based (+17% ($p = 0,0009$)) models. The statistical results using Wilcoxon test validates the significance of the improvements ($p < 0.05$). This again shows the effectiveness of this approach in re-ranking results.

4.4 Comparison with Other Techniques

Query Expansion. Query expansion consists of adding words similar to the original query keywords. We tested various techniques using text-based retrieval results for PRF and some semantic resources based on ontologies such as Word-Net and Babelnet[5] [19].

PRF technique consists in selecting the k top documents of traditional text retrieval results (lucene in our case) that are considered the most relevant. Then, the n most frequent words in this set of documents are added to the initial query for a new search. We tested different values of k (number of selected documents) and n (number of words) where the best configuration, according to statistical analysis of results, is obtained with $k = 5$ and $n = 10$.

For the expansion with WordNet, we evaluate the hyponymy relationships that represent the notion of subclass, and synonyms that are grouped in the form of synsets. We add to the original query terms all synonym terms. These types of semantic relationships can help to extract specific information and terms similar to the query terms.

Finally, we use Babelnet which is a wide coverage multilingual semantic network integrating the WordNet relational structure with the Wikipedia semi-structured information. As in the case of WordNet, each concept is associated with a textual description extracted from WordNet or Wikipedia. Using this semantic resource, we extracted in a first step the set of multi-term phrases (like "bar code", "family tree", etc.) that may occur in the query. Then, we enrich the query by the textual descriptions (Babelnet Gloss) associated to its terms or expressions. Table 4 shows the obtained results.

In these experiments, PRF based expansion outperforms WordNet and Babelnet based expansions. This may be due to the use of terms related to the original query keywords in PRF compared to the use of similar terms but not

---

[5] http://babelnet.org/

Table 4. Evaluation results of query expansion methods

Method	P@5	P@10	MAP
PRF Expansion	0,336	0,308	0,234
Babelnet Gloss	0,216	0,206	0,165
WordNet Hyponym	0,264	0,218	0,197
WordNet Synonym	0,340	0,284	0,229
B + LDA-P (0,7)	**0,400**	**0,336**	**0,281**

related to original query keywords obtained form WordNet or Babelnet Gloss. However, all these expansion methods do not enhance the standard TF-IDF model using lucene similarity score.

Conceptual Retrieval We carry out the extraction of the concepts using Babelnet ontology since it is wide coverage compared to WordNet. Moreover, the fact that this ontology is multilingual allows the extraction of additional concepts in different languages. We apply the process of concept extraction to the queries and the collection of documents similarly. Here an example of queries transformation: satellite image of desert ⇒ "satellite image" "desert", portrait of che guevara ⇒ "portrait" "che guevara". After that we performed CF-IDF (CF for Concept Frequency) weighting scheme for conceptual retrieval using lucene. Table 5 shows obtained results.

Table 5. Results of conceptual retrieval method

Method	Best α Value	P@5	P@10	MAP
Babelnet Concepts	-	0,308	0,262	0,185
B + Babelnet Concepts	0,9	0.380	0.320	0.237
B + LDA-P	0,7	**0,400**	**0,336**	**0,281**

According to these experiments, the use of concepts in image retrieval is not interesting neither in retrieval nor in re-ranking of text-based results. As in the case of topic model, the conceptual representation is reduced in dimensionality compared to the plat textual representation resulting in a loss of precision.

5   Conclusions and Perspectives

We studied in this paper the impact of using LDA topic model in context-based image retrieval by applying it in two granularity levels: global level represented by the entire document, and local level represented by document's paragraphs containing at least one image. We compare our proposed approach with some expansion and conceptual retrieval methods. This work demonstrates the importance of using the local level in image retrieval and image re-ranking tasks. Furthermore, LDA applied to image paragraphs demonstrated a greet success when combined with classical text-based image retrieval. Furthermore, all query expansion approaches does not enhance image retrieval accuracy. Conceptual retrieval using Babelnet concepts is also inadequate for this task.

In future work, we plan to enhance the specific image representation that can be the key for useful retrieval with textual queries. We would like also to investigate how to optimize the selection of suitable words and the rejection of other words. Finally, more work is needed for the LDA topic model especially in the choice of suitable parameters.

References

1. Arora, S., Ge, R., Moitra A.: Learning topic models - Going beyond SVD. In: IEEE 53rd Annual Symposium on Foundations of Computer Science, pp. 1–10 (2012)
2. Barnard, K., Duygulu, P., Forsyth, D.A., de Freitas, N., Blei, D.M., Jordan, M.I.: Matching words and pictures. J. Mach. Learn. Res. **3**, 1107–1135 (2003)
3. Blei, D.M., Jordan, M.I.: Modeling annotated data. SIGIR 2003: Proceedings of the 26th Annual International ACM SIGIR Conference on Research and Development in Information Retrieval, pp. 127–134. ACM (2003)
4. Blei, D.M., Ng, A.Y., Jordan, M.: Latent Dirichlet allocation. J. Mach. Learn. Res. **3**, 993–1022 (2003)
5. Chaney, A.J.B., Blei, D.M.: Visualizing topic models. In: International AAAI Conference on Social Media and Weblogs (2012)
6. Cheng, D., He, X., Liu, Y.: Analyzing the Number of Latent Topics via Spectral Decomposition. arXiv preprint arXiv:1410.6466 (2014)
7. El Demerdash, O., Kosseim, L., Bergler, S.: Image retrieval by inter-media fusion and pseudo-relevance feedback. In: Peters, C., Deselaers, T., Ferro, N., Gonzalo, J., Jones, G.J.F., Kurimo, M., Mandl, T., Peñas, A., Petras, V. (eds.) CLEF 2008. LNCS, vol. 5706, pp. 605–611. Springer, Heidelberg (2009)
8. Griffiths, T., Steyvers, M.: Finding scientific topics. Proc. Natl. Acad. Sci. U.S.A. **101**, 5228–5235 (2004)
9. Gulati, P., Sharma, A.K.: Ontology Driven Query Expansion for Better Image Retrieval. Int. J. Comput. Appl. **5**(10), 33–37 (2010)
10. Harashima, J., Kurohashi, S.: Relevance feedback using latent information. In: Proceedings of the 5th International Joint Conference on Natural Language Processing, Chiang Mai, Thailand, pp. 1037–1045 (2011)
11. Hoffman, M., Blei, D., Cook, P.: Content-based musical similarity computation using the hierarchical Dirichlet process. In: ISMIR 2008–9th International Conference on Music Information Retrieval, pp. 349–354 (2008)
12. Hong, L., Davison, B.D.: Empirical study of topic modeling in twitter. In: Proceedings of the First Workshop on Social Media Analytics, pp. 80–88. ACM (2010)
13. Hörster, E., Lienhart, R., Slaney, M.: Image retrieval on large-scale image databases. In: CIVR 2007: Proceedings of the 6th ACM International Conference on Image and Video Retrieval, pp. 17–24. ACM (2007)
14. Juan, C., Jintao, L., Yongdong, Z., Sheng, T.: LDA-based retrieval framework for semantic news video retrieval. In: International Conference on Semantic Computing. ICSC, IEEE Computer Society, pp. 155–160 (2007)

15. Kullback, S., Leibler, R.: On information and sufficiency. Ann. Math. Stat. **22**(1), 79–86 (1951)
16. Leung, C.H., Li, Y.: Comparison of different ontology-based query expansion algorithms for effective image retrieval. In: Kim, T.-H., Adeli, H., Ramos, C., Kang, B.-H. (eds.) Signal Processing. Image Processing and Pattern Recognition. Springer, Heidelberg (2011)
17. Lu, C., Hu, X., Chen, X., Park, J., He, T., Li, Z.: Probabilistic models for topic learning from images and captions in online biomedical literatures. In: Proceedings of the 18th ACM Conference on Information and Knowledge Management, pp. 495–504 (2009)
18. Maillot, N., Chevallet, J.-P., Valea, V., Lim, J. H.: IPAL Inter-Media Pseudo-Relevance Feedback Approach to ImageCLEF 2006 Photo Retrieval. Working Notes for the CLEF 2006 Workshop (2006)
19. Navigli, R., Ponzetto, S.P.: BabelNet : Building a very large multilingual semantic network. In: Proceedings of the 48th Annual Meeting of the Association for Computational Linguistics. Association for Computational Linguistics, Uppsala, Sweden, pp. 216–225 (2010)
20. Nguyen, C.T., Kaothanthong, N., Phan, X.H., Tokuyama, T.: A feature-word-topic model for image annotation. In: Proceedings of the 19th ACM International Conference on Information and Knowledge Management. ACM, pp. 1481–1484 (2010)
21. Putthividhya, D., Attias, H.T., Nagarajan, S.S.: Supervised topic model for automatic image annotation. In: 2010 IEEE International Conference on Acoustics Speech and Signal Processing, pp. 1894–1897. IEEE (2010)
22. Serizawa, M., Kobayashi, I.: A study on query expansion based on topic distributions of retrieved documents. In: Gelbukh, A. (ed.) CICLing 2013, Part II. LNCS, vol. 7817, pp. 369–379. Springer, Heidelberg (2013)
23. Tang, S., Zheng, Y., Cao, G., Zhang, Y., Li, J.: Ensemble Learning with LDA Topic Models for Visual Concept Detection. In: Multimedia - A Multidisciplinary Approach to Complex, Issues, pp. 175–200 (2012)
24. Teh, Y.W., Newman, D., Welling, M.: A collapsed variational Bayesian inference algorithm for latent Dirichlet allocation. In: Advances in Neural Information Processing systems, pp. 1353–1360 (2006)
25. Troelsgård, R., Jensen, B.S., Hansen, L.K.: A Topic Model Approach to Multi-Modal Similarity. CoRR (2014)
26. Ullah, R., Jaafar, J.: Exploiting short query expansion for images retrieval. International Conference on Computer & Information Science (ICCIS), vol. 1, pp. 352–356. IEEE(2012)
27. Wei, X., Croft, W.B.: LDA-based document models for ad-hoc retrieval. In: Proceedings of the 29th Annual international ACM SIGIR Conference on Research and Development in Information Retrieval, pp. 178–185. ACM (2006)
28. Ye, Z., Huang, X., Lin, H.: Finding a good query-related topic for boosting pseudo-relevance feedback. J. Am. Soc. Inf. Sci. Technol. Arch. **62**(4), 748–760 (2011)
29. Yi, X., Allan, J.: Evaluating topic models for information retrieval. In: Proceedings of the 17th ACM conference on Information and Knowledge management, pp. 1431–1432. ACM (2008)
30. Zhang, M., Luo, C.: A new ranking method based on latent dirichlet allocation. J. Comput. Inf. Syst. **8**(24), 10141–10148 (2012)
31. Zhou, D., Wade, V.: Latent document re-ranking. In: Proceedings of the 2009 Conference on Empirical Methods in Natural Language Processing, vol. 3, pp. 1571–1580. Association for Computational Linguistics (2009)

Social and Semantic Web

Exploiting Microdata Annotations
to Consistently Categorize Product Offers
at Web Scale

Robert Meusel[✉], Anna Primpeli, Christian Meilicke,
Heiko Paulheim, and Christian Bizer

Data and Web Science Group, University of Mannheim, Mannheim, Germany
{robert,anna,christian,heiko,chris}@dwslab.de
http://dws.informatik.uni-mannheim.de

Abstract. Semantically annotated data, using markup languages like
RDFa and Microdata, has become more and more publicly available in the
Web, especially in the area of e-commerce. Thus, a large amount of structured product descriptions are freely available and can be used for various
applications, such as product search or recommendation. However, little
efforts have been made to analyze the *categories* of the available product
descriptions. Although some products have an explicit category assigned,
the categorization schemes vary a lot, as the products originate from thousands of different sites. This heterogeneity makes the use of supervised
methods, which have been proposed by most previous works, hard to apply.
Therefore, in this paper, we explain how *distantly supervised* approaches
can be used to exploit the heterogeneous category information in order to
map the products to set of target categories from an existing product catalogue. Our results show that, even though this task is by far not trivial, we
can reach almost 56 % accuracy for classifying products into 37 categories.

Keywords: Microdata · RDFa · Structured web data · Classification

1 Introduction

Over the last years, more and more websites started making use of markup languages like RDFa, Microformats and Microdata to semantically annotate entities
describing for example events, products, organizations, and persons within their
HTML pages. Those annotations can be parsed, and as they make use of well-
defined vocabularies also interpreted by machines. With the amount of (freely)
available data, this data space becomes more and more interesting in order to
create new knowledge bases, enrich existing knowledge bases, or use the data to
improve applications.

In particular, as recent studies have shown, especially such annotations have
become more and more common in the e-commerce domain.[1] An important step

---

[1] http://webdatacommons.org/structureddata/index.html#toc3.

© Springer International Publishing Switzerland 2015
H. Stuckenschmidt and D. Jannach (Eds.): EC-Web 2015, LNBIP 239, pp. 83–99, 2015.
DOI: 10.1007/978-3-319-27729-5_7

towards exploiting that data is to obtain a detailed profiling of the data that is available. Besides basic information as the number of different instances or the amount of missing information, the category of the products and the distribution of the categories in a dataset is very important.

Although the most common vocabulary used for semantic annotations in the Web, *schema.org*, allows the markup of category information for a product description, the sites across the Web do not make use of one global homogeneous categorization schema. Instead, most sites use their own categorization systems. In order to get a unified view on the categories of products, most previous works made use of supervised methods. Those methods depend on data which is already annotate with the categories from a given target schema. Since a unified set of categories is necessary (and desired), the categories which are annotated cannot be exploit directly for this type of method, and an additional-manual annotation is necessary. As this work is most of the time costly, we propose to use distant supervision as an alternative, using the existing category systems as input. The potential advantages are the reduction of manual work in creating labeled data and keeping the data up-to-date.

The rest of this paper is structured as follows: First, we give a brief introduction to the deployment of (product-related) markup languages and vocabularies in the Web. The next section introduces the corpus we use to evaluate the supervised and distant supervised approaches which can be used to categorize the products. In the following section we state the results for a supervised approach, and then, we explain how our proposed approach, which is based on the idea of distant supervision, can be used to omit the necessity of manually labeled data to train a predictive model. After discussing related work in Sect. 5, we explain the benefits and drawbacks of our proposed method and line out further open challenges in the conclusion.

2   Statistics on Deployed Product-Related Schema.org Microdata in the Web

In this section, we give a brief overview of the available marked up data within the Web. As an object of this analysis, we use the latest extraction of Web-DataCommons (WDC), which includes over 5 billion marked up entities by one of the three main markup languages and has been retrieved from the CommonCrawl corpora of December 2014.[2] From these data, Table 1 describes the four major vocabularies which are used to describe product-related information, together with the specific classes and the number of deploying pay-level domains (PLDs).[3]

---

[2] http://blog.commoncrawl.org/2015/01/december-2014-crawl-archive-available/.

[3] Similar to our previous works [9], we will analysis the data based on PLDs embedding certain vocabularies, classes and properties.

Table 1. Most common product-related deployed classes and vocabularies by number of PLDs in the 2014 corpus.

Major markup format	Vocabulary	Related classes	# of PLDs
Microdata	schema.org	Product, Offer	98 608
Microdata	data-vocabulary.org	Product, Offer	16 003
RDFa	Open Graph Protocol	product	14 592
RDFa	purl.org/goodrelations	Offering	2 196

Good Relations is the original ontology which was later adopted by *schema.org* (`s:Offer` and `s:Product`) to model product-related classes.[4] PLDs (e.g. `swimoutlet.com`, `surveillance-video.com`, and `craftsman.com`) annotating information using the class `gr:Offering` make use of the property `gr:name` in 37 % of the cases, but in 70 % of the cases, they the property `gr:description`. This might be an effect of the sometimes rather small number of crawled pages of non-popular sites.

Open Graph Protocol is mainly used by Facebook to integrate external entities into the Facebook eco-system. From the sites making use of the `product` class in this vocabulary (e.g. `bebe.com` or `epicsports.com`), over 70 % also mark the `title`, `image`, `url`, and `description`.

Data-Vocabulary.org is used within Microdata. It is the predecessor of *schema.org* and is still adopted widely in the Web. Among more than 9 000 PLDs still using the `dv:Product` class in our corpus, we find also well known domains like `samsung.com` and `audible.com`. Similar to `gr:Offering`, only a small fraction (< 50 %) of the PLDs make use of `dv:name`, and only one third annotate a `dv:description`.

Schema.org is the most frequently used vocabulary to describe products. 89 608 PLDs (10.9 %) annotate at least one entity as `s:Product` and 62 849 PLDs (7.6 %) annotate at least one entity as `s:Offer`.

Table 2 shows the number and percentage of the most common used and three selected properties embedded by PLDs making use of the classes `s:Product` and `s:Offer`. Especially for the focus of the paper, the PLDs making use of `s:category` and `s:breadcrumb` are important. Here, we see only a small number of PLDs at all which annotate this information.

From the PLDs making use of *schema.org*, we identified 43 frequently visited e-commerce sites based on the reports by *Bloomberg*[5], and *Yahoo! Finance*[6]

---

[4] http://blog.schema.org/2012/11/good-relations-and-schemaorg.html.
[5] http://www.bloomberg.com/ss/08/11/1107_ecommerce/12.htm.
[6] http://finance.yahoo.com/news/research-markets-worlds-leading-e-154500570. html.

Table 2. Most common used and selected properties by PLDs deploying `s:Product` or `s:Offer`. * marks properties defined for `s:Product`. ** marks properties defined for `s:Offer`. *** marks properties defined for `s:WebPage`

Property	# Product PLDs	% Product PLDs	# Offer PLDs	% Offer PLDs
s:name*	78 292	87.37 %	54 193	86.23 %
s:image*	59 445	66.34 %	45 824	72.91 %
s:description*	58 228	64.98 %	42 730	67.99 %
s:offers*	57 633	64.32 %	55 630	88.51 %
s:price**	54 290	60.59 %	59 452	94.59 %
s:availability**	36 789	41.06 %	37 871	60.26 %
s:priceCurrency**	30 610	34.16 %	32 114	51.10 %
s:url*	23 723	26.47 %	15 601	24.82 %
s:aggregateRating*	21 166	23.62 %	12 325	19.61 %
s:category**	1 479	1.65 %	1 667	2.65 %
s:breadcrumb***	431	0.48 %	460	0.73 %

and the traffic volume of those PLDs (based on Alexa[7]). From those identified PLDs, `shopping.yahoo.com`, `hm.com`, and `oodle.com` were not contained in the original crawl (e.g. due to restrictions within the *robots.txt*). Eight of the remaining 40 embed Microdata, but do not use product-related classes, for example `amazon.com`, which annotates the videos of their instant view platform, but not the physical products. We divided the remaining 32 into the three e-commerce roles: *producer*, *merchant* and *marketplace*, and analyzed the usage of the most common properties (based on all sites making use of products-related markup).

The results of this analysis can be found in Table 3. We found that except of the usage of `s:description` by the identified marketplaces, it is more likely that the identified e-commerce pages make use of the selected properties to annotate

Table 3. Analysis of property usage of selected 32 e-commerce sites.

	Producer	Merchant	Marketplace	Overall
s:name	87.5 %	93.8 %	87.5 %	87.4 %
s:image	75.0 %	75.0 %	62.5 %	66.3 %
s:description	75.0 %	56.3 %	37.5 %	65.0 %
s:offers	75.0 %	81.3 %	62.5 %	64.3 %
s:price	75.0 %	75.0 %	62.5 %	60.6 %
s:priceCurrency	25.0 %	56.3 %	50.0 %	34.2 %
s:availability	25.0 %	56.3 %	50.0 %	41.1 %

---

[7] http://www.alexa.com/.

their products, than all product-related sites in general. In comparison to each other, merchants use the selected properties slightly more often.

Based on these findings, such more frequently visited sites are a good entry point to gather product descriptions with a minimal set of properties.

In the following we explain how the annotated data can be used to assign categories for a given set of products. Therefore, we first introduce the data we use for evaluation and further explain our proposed approach step by step.

3   Experimental Setup

In this section, we describe the data and categorization schema used in our experiments, as well as the gold standard and the evaluation measures used.

Table 4. Product data examples

s:name	s:decription	s:brand	s:category/s:breadcr
ColorBox Stamp Mini Tattoo	ColorBox Stamps are easy to use and perfect for papercraft fun. [..] Not for use by children 12 years andyounger	ColorBox	Stamps >Rubber Stamp
Cowhide Link Belt	ITEM: 9108 Your search is overfor a great casual belt for jeans or khakis. [..]	-	Accessories
Fiesta SE	Automatic, Sedan, I4 1.60L , Gas, RedVIN: 3FADP4BJ8DM1679	Ford	cars
Alabama Crimson Tide Blackout Pullover Hoodie - Black	No amount of chilly weather can keep you from supporting your team.[..]	-	Alabama Crimson Tide >40to60
231117-B21 HP PIII P1266 1.26GHz ML330 G2	Description:Pentium III P1266 ML330 G2/ML350 G2/ML370G2 (1.26GHz/ 133MHz/512K/43W) [..] # 231117-B21	HP Compaq	G2 Xeon
TFS Lil' Giant Anvil, 65 lb	Dimensions: Face 4"x 10.75" Horn 4"x 8.25" Height8"Base 9.25" x 11"Hardie Hole: 1" [..] #: TFS7LG65	Anvils [..]	Hardware >Tools >Anvils
Gavin Road Cycling Shoe	For great performance at adiscounted price, [..]	-	Root RoadBikeOutlet.com >Apparel >Shoes >>

3.1   Product *Schema.org* Microdata

From the whole WDC 2014 Microdata corpus[8], we derived a subset of 9 414 product descriptions from 818 different PLDs. We have chosen products from PLDs for which each product description uses at least the properties s:name, s:description, and s:brand, and either one of the two properties s:category (84 % of the PLDs) or s:breadcrumb (16 % of the PLDs). From each PLD, we extracted at most 20 products to reduce the risk of a bias towards a certain category. Table 4 shows an excerpt of the data. Especially for the categories/breadcrumb values, we observed a mixture of multi-level and flat paths, as well as tag-like annotations. 3 653 respectively 1 019 distinct s:category values respectively s:breadcrumb values are used by the included products.

3.2   GS1 - Global Product Catalogue

For our experiments, we used the *GS1 Product Catalogue* (GPC) as target hierarchy. The GPC is available in different languages and claims to be a standard for everything related to products.[9] The hierarchy is structured in six different levels starting from the *Segment*, over *Family*, *Class*, and *Brick*, down to the last two levels *Core Attribute Type* and *Core Attribute Value*. The first level distinguishes between 38 different categories, the second level divides the hierarchy into further 113 categories and the third level consists of 783 disjunct categories. In addition to the textual labels for each category in the hierarchy, the forth and the sixth level partly include a – more or less – comprehensive textual description. Table 5 shows the first four levels of three paths of the hierarchy.

Table 5. Excerpt of GS1 GPC (first four levels). [..] is a placeholder, if the label is similar to the one of the former level.

Segment	Family	Class	Brick
Toys/Games	[..]	Board Games/Cards/Puzzles	Board Games (Non Powered)
Food/Beverage/Tobacco	Seafood	Fish Prepared/Processed	[..] (Perishable)
Footwear	[..]	Footwear Accessories	Shoe Cleaning

3.3   Gold Standard

Using the set of categories from the previously mentioned hierarchy, we manually annotated the set of products described in Sect. 3.1. Specifically, we annotated each product (if possible) with one category for each of the first three levels. The annotations were performed by two independent individuals. Whenever there was a conflict, a third person was asked to solve the discrepancy. The annotators were first asked to read the title/name, description, and the additional available values

---

[8] http://webdatacommons.org/structureddata/2014-12/stats/schema_org_subsets.html.
[9] http://www.gs1.org/gpc.

of the product and, in case of insufficient information, they should visit the web page of the product.

Within the gold standard, we could not assign any category to 187 (2.09 %) products, mostly because the available attributes to describe the products were insufficient, and the web page of the product was either not available any more or here also not enough information were given. Based on the first level of the GS1 GPC hierarchy, we assigned at least each category once (except *Cross Segment*). Table 6 depicts the ten most frequent categories of the first level within the gold standard. We see a domination by the category *Clothing*. For the second level, we assigned 77 (68.14 %) different labels at least once, and 303 (38.70 %) different labels for the third level. The gold standard, as well as more comprehensive statistics, can be found at our website.[10]

Table 6. Distribution of categories for the first Level of the GS1 GPS within the gold standard dataset, as well as with the predicted distributions of the best supervised and distant supervised approach.

Rank	Category level 1	Original	Supervised	Δ	Distant superv	Δ
1	Clothing	0.435	0.401	0.033	0.406	0.028
2	Personal Accessories	0.053	0.128	0.075	0.039	0.014
3	Household/Office Furniture/Furnishings	0.051	0.045	0.006	0.035	0.016
4	Automotive	0.047	0.054	0.007	0.052	0.005
5	Computing	0.037	0.034	0.004	0.023	0.014
6	Audio Visual/Photography	0.036	0.030	0.006	0.020	0.015
7	Healthcare	0.033	0.027	0.006	0.005	0.027
8	Pet Care/Food	0.026	0.028	0.002	0.017	0.010
9	Sports Equipment	0.026	0.030	0.004	0.022	0.004
10	Food/Beverage/Tobacco	0.024	0.025	0.001	0.007	0.018
11-38	*Others*	0.232	0.198	0.065	0.373	0.159

3.4 Baseline and Evaluation

As we want to show to which extent the categorizations of the single PLDs can be used to assign categories from a global hierarchy to products, we compare ourselves to the results of a supervised classification approach. The approach is trained using 10-fold cross-validation with three different classification methods: Naive Bayes (NB), Decision Trees (DT) and k-Nearest Neighbor approach, where $k = 5$ (5-NN). A detailed description of the baseline method can be found in Sect. 4.2.

---

[10] http://webdatacommons.org/structureddata/2014-12/products/gs.html.

For reasons of comparison, we use *accuracy* (ACC) as the main evaluation metric. Whenever an approach is not able to return a category for a given product, we count this examples as a *false negative*. For approaches returning either one label or no label for each instance, this measure is equal to recall (R). In addition, for our distant-supervised approaches, we also report the precision P, as this measure gives an idea about the performance of predicted labels, without regard of those which cannot be labeled. We also state the f-score $F1$, representing a trade-off between R and P.

4   Experiments and Results

In this section, we will first state how the both input data sources, i.e., product descriptions and categories from a given target hierarchy, are transformed into feature vectors that can be processed by further methods. Then, we train a model based on the hand-annotated categories of the gold standard. The remaining parts of this section introduce our distant supervision approach making use of the categorical information for the products given on the PLDs itself.

4.1   Feature Vector Generation

As stated above, we have two types of input: products, which are described by a set of properties, and the categories of the target hierarchy. In order to transform both types of input into comparable feature vectors, we generate a *bag of words* representation for each entity, i.e., each product and each category at a certain depth within the hierarchy.

For the products, we experiment with different sets of property combinations (e.g. only `s:title`, `s:title` with `s:description`, and so on). For the hierarchies, we use the textual names of the categories themselves and all or a selection of the names of sub-categories (e.g., segment, segment and family, segment and brick). In all cases, we tokenize the textual values by non alpha-numeric characters, remove stopwords and stem the data using a *Porter Stemmer*. Moreover, we transform all characters to lower case and remove terms which are shorter than 3 and longer than 25 characters.

In order to weight the different features for each of the elements in the two input sets, we apply two different strategies:

Binary Term Occurrence (BTO), where the weight of a term for an element is either 1, if the term occurs at least once within the textual attributes of the element, 0 otherwise.

TF-IDF, where the term frequency is normalized by the inverse document frequency, which removes the impact of common terms which occur in a large fraction of documents.

In the following, we refer to the set of feature vectors describing products by *Pro* and to those describing labels of the categories of the hierarchy by *Cat*. Depending on the textual attributes which were used to create the vectors, the number of final attributes ranges between 4 000 (only category and breadcrumb) to around 11 000 (all properties).

4.2 Baseline: Supervised Approach

Table 7 presents the results with different setups for the baseline classification approach. We reach the highest accuracy with a 5-NN classification algorithm using *Jaccard Coeffcient*. Decision Trees did not perform at a comparable level, so we excluded them from the table. We also calculated the distribution of the predicted product categories for the best approach. The results are shown in Table 6 including the deviation from the distribution of categories in the gold.

Table 7. Selected results of the baseline classification for assigning GS1 GPC first level categories. Highest scores are marked in **bold**.

Selected properties	Term weight	Classifier	ACC
Name,Desc	BTO	NB	.722
Name,Description	TF-IDF	NB	.733
Name,Description	BTO	5-NN(Jaccard)	.608
Name,Description	TF-IDF	5-NN(Cosine)	.728
Name,Description,Categroy,Breadcr.	BTO	NB	.754
Name,Description,Categroy,Breadcr.	TF-IDF	NB	.757
Name,Description,Categroy,Breadcr.	BTO	5-NN(Jaccard)	.819
Name,Description,Categroy,Breadcr.	TF-IDF	5-NN(Cosine)	.740
Name,Description,Categroy,Breadcr.,Brand	BTO	NB	.758
Name,Description,Categroy,Breadcr.,Brand	TF-IDF	NB	.760
Name,Description,Categroy,Breadcr.,Brand	BTO	5-NN(Jaccard)	**.820**
Name,Description,Categroy,Breadcr.,Brand	TF-IDF	5-NN(Cosine)	.746

4.3 Hierarchy-Based Product Classification

In a first step, we use the feature vectors created for the categories from the target hierarchy *Cat* in order to train a predictive model (one labelled example for each category). This model is then used to predict the labels for the instances of *Pro*. We test different classification methods, namely *Naive Bayes* (NB), *k-Nearest-Neighbour* with $k = 1$ (1−NN)[11], *Support Vector Machines* (SVM), and *Random Forests* (RF).

Table 8 shows the results of the best configuration, using only the features from the values of the properties name, category and breadcrumb from *Pro* and all hierarchy labels from the GS1 GPC. We find that on average TF-IDF as term weighting methods performs better than a BTO strategy. The best results are achieved using 1-NN and Naive Bayes classification on TF-IDF vectors.

---

[11] As for each class, only one example exists k needs to be set to 1, otherwise the method would consider other examples then the nearest, which by design belong to another class. This setup is equal to *Nearest Centroid Classification*, where each feature vector of *Cat* is equal to one centroid.

Table 8. Best results achieved with distant supervised classification using instances of *Cat* for training. Highest scores are marked in **bold**.

Term weighting	Classifier	ACC
TF-IDF	NB	**.377**
TF-IDF	1-NN (Cosine)	**.377**
TF-IDF	1-NN (Jaccard)	.361
TF-IDF	SVM	.376
TF-IDF	Random Forest	.006
BTO	NB	.000
BTO	1-NN (Cosine)	.330
BTO	1-NN (Jaccard)	.271
BTO	SVM	.000
BTO	Random Forest	.026

4.4   Similarity-Based Product Category Matching

In order to exploit the promising performance of the distance-based classification approach (1-NN) of the former section, we extend our approach in this direction, using the similar fundamental idea as nearest-neighbour classifier. We calculate for each instance in *Pro* the distance to all instances in *Cat*. To that end, we use three different similarity functions, namely:

Cosine Similarity: This measure sums up the product of the weights/values for each attribute of the two vectors and is supposed to work well with TF-IDF.
Jaccard Coefficient: This measure calculates the overlap of terms occurring in both vectors and normalize it by the union of terms occurring in both vectors. This measure is supposed to work well with binary weights.
Non-normalized Jaccard Coefficient: As the descriptions of products could be rather comprehensive (based on the way the data was annotated), we address the penalization of longer product names, which would occur for Jaccard, by introducing a non-normalized version of the *Jaccard-Coefficent*, i.e., only measuring the overlap.

In addition, we use different sets of textual attributes from the products as well as from the hierarchies to build the feature vectors. Based on the similarity matrix, we then select for each instance in *Pro* the instance in *Cat* with the highest score, larger than 0. In contrast to a classifier, we do not assume any distribution within the data, or assign any category randomly. This means in case of two or more possible categories which could be assigned, we do not assign a particular instance from *Cat* to the instance of *Pro*.[12]

Table 9 reports a selection of results of this approach trying to predict the categories of the first, second and third level within the hierarchy. In each of the

---
[12] As stated before, such instances are counted as *false negatives* within the evaluation.

three blocks, the first line always reports the best results using only the category and breadcrumb as input for the feature vector. The second line reports the result for the default configuration (all attributes, TF-IDF). The third line shows the result for the optimized setup of attributes and term weighting. In all cases in the table cosine similarity produced the best results. In some experiments the other two similarity functions performed comparable, but overall did not produce better results. Starting from level one to three, we see a slight decrease in terms of accuracy. This is not surprising as the number of possible labels increases with each level (see Sect. 3.3) and the contentual boundaries between them become more and more fuzzy. In addition, we found that the best configuration just differs by some percentage points from the default configuration for all three levels (e.g. .341 vs. .359 for the first level). Furthermore, using only the information from the category and the breadcrumb alone does not produce the highest accuracy results. For all three levels the best results in terms of accuracy could be reached using the textual values of category, breadcrumb and name as input for the feature vector creation.

Inspecting the results of the optimal solution for each level manually, we found that in most cases the overlap in features between the instances of *Pro* and *Cat* was insufficient for those instances which were wrongly categorized or left unlabelled. Reasons for this are the use of a different language as the target hierarchy (e.g. Spanish), a different granularity (e.g. *fruits* versus *cherry*) or the use of synonyms (e.g. *hat* versus *cap*).

A common method to overcome at least the two latter discrepancies is the enhancement with a external/additional ground knowledge.[13] For our experiments, we use two different sources of ground knowledge to enhance our feature vectors. First, we make use of the *Google Product Catalogue*.[14] This catalog is used by Google to enable advertisers to easily categorize the ads they want to place. The catalog is available in different languages and in addition is more tailored towards products traded in the Web. The second source we use is based on the co-occurrences of different terms within a corpus. In particular, we make use of *extracting DIStributionally related words using CO-occurrences* framework (DISCO)[15], first presented by Kolb [5], where we load the English Wikipedia and enhance the feature vectors of the categories.

The best results and the comparison to the best results without the enhancement can also be seen in Table 9, within the third and forth row of each block. In general we find a strong increase in the accuracy in comparison to the non-enhanced experiments. For the first level, we increase our performance by 33 % to almost .5 accuracy. For level three, however, this effect diminishes almost completely. Even with the enhanced vectors, the improvements are small.

In the following we describe two different types of experiments to further improve our results. In the first, we concentrate on high-precision results and

---

[13] We thank Stefano Faralli for his valuable feedback and recommendations.
[14] https://support.google.com/merchants/answer/1705911?hl=en.
[15] http://www.linguatools.de/disco/disco_en.html.

Table 9. Selected results for all three category levels, including the default configuration, the best with and without ground knowledge.

Level	Product properties	Product weight	Hierarchy term levels	Hierarchy term weight	Ground knowledge	ACC	P	F1
1	Category, Breadcr	BTO	1-6	TF-IDF	none	0.288	0.334	0.309
1	All	TF-IDF	1-6	TF-IDF	none	0.341	0.344	0.343
1	Name, Category, Breadcr	BTO	1-6	TF-IDF	none	0.359	0.373	0.366
1	Name, Category, Breadcr	BTO	1-4	TF-IDF	DISCO	0.479	0.499	0.489
2	Category, Breadcr	BTO	1-6	TF-IDF	none	0.171	0.297	0.217
2	All	TF-IDF	1-6	TF-IDF	none	0.261	0.264	0.263
2	Name, Category, Breadcr	BTO	1-6	TF-IDF	none	0.294	0.305	0.300
2	Name, Category, Breadcr	BTO	1-4	TF-IDF	DISCO	0.380	0.395	0.387
3	Category, Breadcr	BTO	1-6	TF-IDF	none	0.109	0.112	0.111
3	All	TF-IDF	1-6	TF-IDF	none	0.196	0.198	0.197
3	Name, Category, Breadcr	BTO	1-6	TF-IDF	none	0.257	0.267	0.262
3	Name, Category, Breadcr	BTO	1-4	TF-IDF	DISCO	0.258	0.269	0.263

obtained those values as labeled instances. Then, we train a predictive model on those instances. In the second approach, we reformulate the task of labeling a set of instances as a global optimization problem.

4.5   Classification on High-Precision Mappings

This approach is based on the idea that, even if the accuracy (which represents the global performance of the matching) is not sufficient, we could make use of those instances which were assigned to a category with a high precision. Those instances can further be used as input in order to train a predictive model. It is important to note that when selecting the mapping, all, or at least a large fraction of categories (which should be predicted), should be included. This means that some configurations even with $P = 1$ are not useful, as they include too few instances. In order to improve the precision of our initial mapping, we introduce a higher minimal similarity between products and categories.

The first columns in Table 10 show the highest precisions which could be reached, where at least 100 product instances were assigned to an instance of Cat of level 1. The precision of those optimal configurations ranges between .75 and .79, which means that within this data, every fourth or fifth instance is wrongly labeled. In addition, we report the values for a setup with less precision (.57) but with over 5 500 labeled examples.

We tested different mappings and train different classification methods on this input data. In Table 10 we outline the best performing results for the different input configurations.[16]

Table 10. Result of combined approach, using high-precision mapping result as training data to learn a predictive model for level 1 categorization.

Product Properties	Product Weight.	Hierarchy Levels	Term Weight.	Ground Knowldg.	Min. Sim.	Mapping ACC	P	# Inst.	Classifier	Overall ACC
Name,Cat.,Breadcr.	BTO	1-6	TF-IDF	Google	>.35	0.009	0.789	109	NB	0.076
All	BTO	1-6	TF-IDF	Google	>.25	0.008	0.772	103	5-NN	0.079
Name,Cat.,Breadcr.	TF-IDF	1-6	TF-IDF	Google	>.25	0.028	0.747	340	NB	0.069
Name,Cat.,Breadcr.	BTO	1-4	TF-IDF	Disco	>.05	0.340	0.570	5 505	RF	0.514

We found, that in case of the high-precision configurations (first three rows) the overall precision of the classifier which can be trained based on those input data is poor, and in all three cases did not exceed 10 % accuracy. Manually inspecting those datasets and the resulting classifications reveals that not all classes are included in those sets, so the model cannot predict all classes (as they are unknown) and that the number of training data is not enough even for the classes which are included. Inspecting the results of the fourth configuration, where the final accuracy exceeds slightly the 50 %, we found almost a balanced distribution in the errors of the classification.

4.6   Global Optimization

In the approaches so far, we evaluate each match between an instance in Pro and Cat in isolation. However, the similarity between two products should be used as an indicator for mapping these instances to the same category, and vice versa. Deciding about the similarity of products and matching them to categories are thus highly dependent problems.

We try to take these dependencies into account by formalizing the problem as a global optimization problem in terms of Markov Logic [2]. In particular, we use the solver *RockIt* by Noessner et al. [13] to compute the most probable solution,

---

[16] We also applied up-sampling of under-represented classes in the dataset, but the results did not improve.

also known as the MAP (maximum a posteriori) state. In our formalization, we use two weighted predicates *map* and *sim* to model the mapping of a product to a category and to model the similarity of two products. We use the similarity matrices from the former experiments as input for defining the prior weights for the respective atoms. Then we define the score which needs to be optimized as the sum the weights attached to these atoms. Further, we define two hard constraints which have to be respected by any valid solution. (1) If two products are similar they have to mapped to the same category. (2) Each product can be assigned to only one category.

Using the best configuration of the former similarity-based matching results from Sect. 4.4, where we reached an accuracy level of .479, we tested different combinations for the similarity of products, as well as the minimal similarity we handed into the global optimization problem. In addition, we also tested different weight ratios between the two predicates, where we multiply the original weight of *map* with a constant factor. In Table 11, we report the best configurations and the corresponding accuracy values. In comparison to the original value of .479 we could improve up to .555, and we assume that this is not the best value which can be reached. Unfortunately, even if running the solver on a large machine, with 24 cores and over $300GB$ of RAM, further experiments did not finish within 24 hours, which shows that the approach is promising, but requires more tweaks to run at large scale.

We have selected the best performing distant supervised approach and calculated again the resulting distribution of product categories (see Table 6). Note that the supervised approach has a summed absolute error of .20 while the best distant supervised approach has a summed absolute error of .31 (the average absolute error is .006 respective .008).

Table 11. Results of the best configurations for solving the optimization problem. Highest scores are marked in **bold**.

| Similarity | | min. value | Weight ratio | | | |
map	sim	for sim	map/sim	ACC	P	$F1$
Cosine	Cosine	0.5	20/1	0.505	0.540	0.522
Cosine	Jaccard	0.5	20/1	0.483	0.506	0.494
Cosine	Cosine	0.5	10/1	0.514	0.556	0.534
Cosine	Jaccard	0.5	10/1	0.484	0.509	0.496
Cosine	Cosine	0.4	10/1	0.553	0.606	0.578
Cosine	Cosine	0.3	10/1	**0.555**	0.636	0.593

5   Related Work

In this section, we describe relevant works both in the area of analyzing the deployment of structured web data in general, as well as in automatic product classification.

5.1   Deployment of Structured Data

The deployment of structured data was first presented by Mika and Potter [10,11] and later an updated view on the adoption of schema.org was given by Guha [4], where [14] analyzed this vocabulary on schema level. In our previous works we analyzed the deployment of the three markup languages in a general matter [1,9], and in addition analyzed the kinds of errors included in such kind of data [8]. Besides, we also inspected how the deployment changes over time [7].

In addition to those markup languages, recent works try to leverage information embedded in HTML tables [3,6,17].

5.2   Categorization of Product Data

Learning a classification model to predict labels for unclassified products was presented in [15]. The authors made use of products and their categories retrieved from amazon.com. Our proposed approaches aims at removing the dependency on external data classification providers.

A recent approach by Qiu et al. [16] presented a system which efficiently detects product specifications from product detail pages for a given category. In order to determine the category, they make use of pre-trained classifiers and a set of seed product identifiers of products related to this category.

Nguyen et al. [12] present an end-to-end solution facing the problem of keeping an existing product-catalogue with several categories and sub-categories up-to-date. Their approach includes data extraction, schema reconciliation, and data fusion. The authors show that they can automatically generate a large set of product specifications and identify new products.

The three mentioned approaches make use of hand-labeled or pre-annotated data, which is not (easily) accessible in larger quantities. This underlines the need of alternative methods to overcome the need of labeled data.

6   Conclusion

In this paper, we have first given a short overview about the deployment of product-related markup languages and vocabularies within HTML pages. In the second part, we have described a subset of this data, which we manually annotated with the categories of the first three levels of the GS1 Global Product Catalogue. Based on that gold standard, we have shown that using supervised methods can reach an accuracy of 80 % when learning a predictive model in order to categorize products.

Further, as already some sites mark products with a site-specific category, we first have shown that using this information alone, due to its heterogeneity among different sites, is not an optimal input for a distantly supervised approach. But in combination with other properties (e.g. the name), that information can be leveraged by distantly supervised methods and thereby assign categories from a given set to products with an accuracy of up to 56 %. To that end, we use various refinements of the problem, taking both background knowledge into account, as well as modeling the categorization of a set of instances as a global optimization problem. The latter provides very promising results, but also hints at scalability issues of solving such optimization problems.

Regarding the distributions which are predicted by the two different kinds of approaches, we see that the supervision works slightly better, but both results can be used in order to gain first insights in the category distribution of the dataset.

Another area where further improvements can be made is the selection of sources. In our gold standard, we only included product descriptions from less than 1 000 PLDs, while on the Web, there are by far more which can be exploited. In particular it might be a promising approach to weight the influence of products of a particular PLD by other attributes, for example the average length of the description or the depth of the given category information.

References

1. Bizer, C., Eckert, K., Meusel, R., Mühleisen, H., Schuhmacher, M., Völker, J.: Deployment of RDFa, microdata, and microformats on the web – a quantitative analysis. In: Alani, H., et al. (eds.) ISWC 2013, Part II. LNCS, vol. 8219, pp. 17–32. Springer, Heidelberg (2013)
2. Domingos, P., Lowd, D.: Markov logic: An interface layer for artificial intelligence. Synth. Lect. Artif. Intell. Mach. Learn. **3**(1), 1–155 (2009)
3. Eberius, J., Thiele, M., Braunschweig, K., Lehner, W.: Top-k entity augmentation using consistent set covering. In: SSDBM 2015 (2015)
4. Guha, R.V.: Schema.org update. http://events.linkeddata.org/ldow2014/slides/ldow2014_keynote_guha_schema_org.pdf, April 2014
5. Kolb, P.: Disco: A multilingual database of distributionally similar words.In: Proceedings of KONVENS (2008)
6. Lehmberg, O., Ritze, D., Ristoski, P., Meusel, R., Paulheim, H., Bizer, C.: Mannheim Search Join Engine. Science, Services and Agents on the World Wide Web, Web Semantics (2015)
7. Meusel, R., Bizer, C., Paulheim, H.: A web-scale study of the adoption and evolution of the schema.org vocabulary over time. In: Proceedings WIMS 2015, pp. 15:1–15:11. ACM, New York, NY, USA (2015)
8. Meusel, R., Paulheim, H.: Heuristics for fixing errors in deployed schema.org microdata. In: Extended Semantic Web Conference (2015)
9. Meusel, R., Petrovski, P., Bizer, C.: The webdatacommons microdata, RDFa and microformat dataset series. In: Mika, P., et al. (eds.) ISWC 2014, Part I. LNCS, vol. 8796, pp. 277–292. Springer, Heidelberg (2014)
10. Mika, P.: Microformats and RDFa deployment across the Web (2011). http://tripletalk.wordpress.com/2011/01/25/rdfa-deployment-across-the-web/
11. Mika, P., Potter, T.: Metadata statistics for a large web corpus. In: LDOW 2012, CEUR Workshop Proceedings, vol. 937. CEUR-ws.org (2012)
12. Nguyen, H., Fuxman, A., Paparizos, S., Freire, J., Agrawal, R.: Synthesizing products for online catalogs. Proc. VLDB Endow. **4**(7), 409–418 (2011)
13. Noessner, J., Niepert, M., Stuckenschmidt, H.: Rockit: Exploiting parallelism and symmetry for MAP inference in statistical relational models. In: Proceedings of the AAAI 2013 (2013)

14. Patel-Schneider, P.F.: Analyzing schema.org. In: Mika, P., et al. (eds.) ISWC 2014, Part I. LNCS, vol. 8796, pp. 261–276. Springer, Heidelberg (2014)
15. Petrovski, P., Bryl, V., Bizer, C.: Integrating product data from websites offering microdata markup. In: DEOS 2014 (2014)
16. Qiu, D., Barbosa, L., Dong, X.L., Shen, Y., Srivastava, D.: Dexter: Large-scale discovery and extraction of product specifications on the web. Proc. VLDB Endowment 8(13), 2194–2205 (2015)
17. Ritze, D., Lehmberg, O., Bizer, C.: Matching html tables to dbpedia. In: Proceedings of the 5th International Conference on Web Intelligence, Mining and Semantics, p. 10. ACM (2015)

The Interactive Effect of Review Rating and Text Sentiment on Review Helpfulness

Shasha Zhou[1] and Bin Guo[1,2(✉)]

[1] School of Management, Zhejiang University, Hangzhou, China
zss_1224@163.com, guob@zju.edu.cn
[2] School of Business, Zhejiang University City College, Hangzhou, China

Abstract. Review ratings and text sentiments respectively represent quantitative and qualitative aspects of user-generated product reviews. These two types of polarity information complement each other in affecting consumers' review evaluation. Few extant studies consider the interplay of review rating and text sentiment on perceived review helpfulness. In this study, we attempt to investigate this potential interaction effect and examine whether it is conditional on review length. The empirical results from an analysis of 70,610 restaurant reviews from Yelp.com indicate that both review ratings and text sentiments exhibit negativity bias effect, such that negative ratings and texts are more helpful than positive ones. Meanwhile, the two types of review valence have a positive interaction effect on perceived review helpfulness. Moreover, the interaction effect of review rating and text sentiment is stronger for longer reviews.

Keywords: Word of mouth · Sentiment analysis · Review helpfulness

1 Introduction

As an important information source for consumers, online product reviews—a form of electronic word of mouth (eWOM), are playing an increasingly important role in the popularity of electronic commerce [1]. It is well recognized that user-generated product reviews significantly affect consumer purchase decisions and product sales [2–6]. However, not all reviews incur the same effect on consumer behavior. The question of why some reviews are perceived to be more helpful and influential than others has received increasing attention within e-commerce research. Various review characteristics, such as review valence, extremity, length, readability, subjectivity, message style, and semantic characteristics, have been found to affect review helpfulness [1, 7–12]. The important roles of reviewer characteristics, such as reviewer identity disclosure, reviewer innovativeness, reviewer expertise and reputation, have also been emphasized in shaping consumers' perception of review helpfulness [4, 11, 13].

An important finding in previous studies is that negative reviews are perceived to be more helpful than positive reviews [14–16], known as negativity bias [17]. When measuring review valence, extant research generally assumed that numeric ratings represent the overall sentiment orientations in review texts. In other words, the valences between numerical ratings and textual contents are consistent, thereby using numeric

© Springer International Publishing Switzerland 2015
H. Stuckenschmidt and D. Jannach (Eds.): EC-Web 2015, LNBIP 239, pp. 100–111, 2016.
DOI: 10.1007/978-3-319-27729-5_8

rating as a proxy measure for review valence. However, this assumption may not always be true, because ratings may not fully capture the polarity of textual information [18]. Sentiments expressed in the texts provide more tacit and context-specific emotions of the reviewers, beyond the numeric ratings [19]. As consumers read the texts of online product reviews rather than rely simply upon summary statistics such as numerical ratings [2], review rating and text sentiment may jointly affect perceived review helpfulness. However, this joint impact of review rating and text sentiment on review helpfulness remains largely unexplored in previous research. This paper is thus motivated to fill this research gap. Specifically, we begin with investigating the impacts of review rating and text sentiment on review helpfulness. Then, we focus on understanding the potential interaction effect of review rating and text sentiment on consumers' perception of review helpfulness. In other words, we are interested in whether the two types of valence strengthen or weaken each other's impact and how they might jointly influence review helpfulness. When review rating is consistent with text sentiment, this consistent rating-text review is expected to be more helpful than when the two types of valence are inconsistent.

Furthermore, we attempt to specify the boundary conditions for the interaction effect of review rating and text sentiment. We are interested in understanding whether this interaction effect is moderated by review length. Longer reviews, on the one hand, may provide more product information and thus attract more consumers' attention and further enhance their expectation. On the other hand, longer reviews require more effort in information processing. As a result, the negative effect of the inconsistency between review rating and text sentiment, which disconfirms expectation and do not reward the information processing effort, may be exacerbated. Hence, we expect a stronger interactive effect of review rating and text sentiment on review helpfulness for long reviews than for short reviews.

Our empirical analysis is based on a rich data set of user-generated restaurant reviews extracted from Yelp.com. We discover that both review rating and text sentiment negatively affect review helpfulness. Meanwhile, review rating and text sentiment have a positive interaction effect on perceived review helpfulness. In addition, we find that this interaction effect is stronger for longer reviews.

The rest of the paper is organized as follows. The next section introduces our theoretical framework and develops our hypotheses. Section three describes our source of data and the empirical model. Main findings are presented and discussed in section four. The paper ends with a discussion of theoretical and managerial implications and suggestions for future research.

2 Theory and Hypotheses

Figure 1 presents our theoretical model, which illustrates the proposed effects of review rating (H1), text sentiment (H2), along with their interaction on consumers' perception of review helpfulness (H3), as well as the moderating effect of review length on the interaction of review rating and text sentiment (H4). Each hypothesis is elucidated in the following subsections.

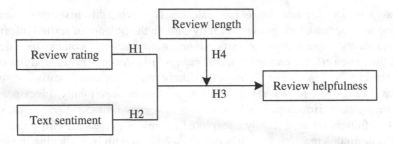

Fig. 1. Research model

2.1 Effects of Review Rating and Text Sentiment

The first issue we explore involves investigating the impacts of review rating and text sentiment on review helpfulness. The extant literature has consistently shown an asymmetric effect of review valence. Specifically, negative reviews are perceived to be more helpful than positive reviews [14, 15]. This negativity bias effect has been explained from different perspectives. One important explanation is that negative reviews are more diagnostic than positive ones [13, 15]. However, much work is based on the impact of review rating. The literature has provided few empirical insights into whether the valence of review text has an asymmetric effect. Specifically, we expect that the negativity bias effect may apply to text sentiment. In other words, reviews with negative sentiments are expected to be more helpful than those with positive sentiments. We propose the following hypotheses:

H1:   Reviews with negative ratings are perceived to be more helpful than reviews with positive ratings.

H2:   Reviews with negative text sentiments are perceived to be more helpful than reviews with positive text sentiments.

2.2 Interaction Effect of Review Rating and Text Sentiment

The second issue we explore is the possible interaction effect between review rating and text sentiment. Review ratings and text sentiments respectively represent quantitative and qualitative aspects of user-generated product reviews. However, these two types of valence may not always be consistent [19, 20]. Considering that consumers look not only at the numerical ratings but also at the textual content of the reviews [2], we expect review rating and text sentiment to jointly affect consumers' perceived review helpfulness. When the valence of review rating is consistent with the valence of review text, the review is expected to be more helpful than those with inconsistent rating-text valences. This is because consistent rating-text valences can enhance a review's trustworthiness [20], while lacking consistency between review rating and text sentiment may make the reviewer appear less capable to provide persuasive information [9]. Moreover, such inconsistency may also engender the difficulty in processing review information. This line of reasoning suggests that the impact of review rating on review helpfulness depends on the sentiment orientation of review text. Hence, we propose the following hypothesis:

H3:   Review rating and text sentiment have an interactive effect on review helpfulness.

2.3   Moderating Effect of Review Length on the Interaction of Review Rating and Text Sentiment

To further specify the boundary conditions for the interaction effect of review rating and text sentiment, the role of review length is particularly considered in this study. We examine whether the rating-text interplay may be contingent on the length of a review. Review length signals the amount of textual content in a review [8]. In comparison with short reviews, long reviews contain more context-specific, product-related information which can help consumers familiarize with the product. Thus, for consumers who seek information from product reviews to reduce uncertainty about product quality, long reviews are likely to attract more attention from consumers than short reviews [21]. The more attention that consumers pay to long reviews may further enhance their involvement in scrutinizing the textual contents, which may improve the comparison between numerical ratings and text sentiments. In this case, the negative effect of inconsistency between review rating and text sentiment will be exacerbated, because the inconsistent information does not reward the information processing effort con-sumers invest in processing long reviews. In addition, more attention and high involvement in long reviews may engender consumers' high expectation for obtaining more helpful and persuasive information. The disparity between expectation and rating-text inconsistency may greatly reduce consumers' review-quality evaluations [22]. Taken together, we expect that the interactive effect may be stronger for long reviews than for short reviews. The following hypothesis is provided.

H4: The interactive effect of review rating and text sentiment on review helpfulness is stronger for long reviews than for short reviews.

3   Research Method

3.1   Sample and Data Collection

We conducted our empirical analysis using a data set of restaurant reviews collected from Yelp.com. We selected Yelp.com because it is one of the most popular product review sites with a large member base and millions of reviews, which offers a rich setting for research on online consumer reviews. Following the sampling strategy of Chen and Lurie [23], we randomly selected 20 restaurants from each of the five major cities including Atlanta, Chicago, Los Angeles, New York, and Washington. Then we extracted all available reviews for each of the 100 restaurants as of December, 10, 2013. For each review, review rating, review date, review text, whether the review text contained picture(s), and the number of "useful" votes a review received were extracted. In addition, the generic information about the reviewers, such as whether (s)he provided a profile photo, the number of friends they had on Yelp, the number of reviews they had ever posted, and whether (s)he was an Elite member, were also extracted. Figure 2 shows an example of a Yelp review and illustrates the information we extracted.

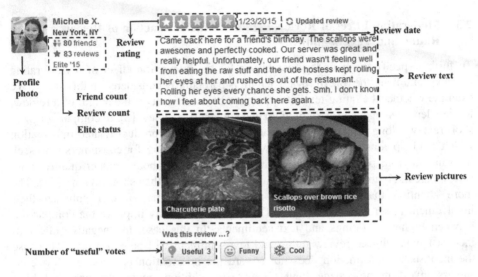

Fig. 2. An example of a Yelp review

In total, 70,610 reviews were obtained. In our sample, 85.2 % of the ratings are positive (4 or 5 stars). This confirms the fact that positive reviews are more prevalent [20]. The disproportionate number of positive ratings is consistent with what prior researchers have found in other online review systems (e.g., Amazon) [13]. In addition, the average review in our sample has 110 words, suggesting more specific information in the review text that a review rating may not capture.

3.2 Measures

Review Helpfulness. In line with prior research (e.g., [23]), review helpfulness was measured by the number of "useful" votes a review received.

Review Rating. Review rating was operationalized using the star rating from 1 (negative) to 5 (positive).

Text Sentiment. The sentiment analysis was automatically performed using the linguistic inquiry and word count (LIWC) program [24]. Following Ludwig et al. [25], the sentiment score for each review was derived through the following formula:

$$sentiment_i = \frac{PW_i - NW_i}{SW_i} \tag{1}$$

where *sentiment$_i$* represents the overall intensity of text sentiment in review *i*; *PW$_i$* is the number of positive content words in review *i*; *NW$_i$* is the number of negative content words in review *i*; *SW$_i$* represents the number of all words in review *i*. According to Eq. (1), the sentiment score obtained from each review ranges from -1 to 1. To facilitate comparison between review rating and text sentiment, we rescaled *sentiment* to a 1 to 5 scale. The sentiment score for each review was recalculated as (max′-min′) × [(v-min)/(max-min)] + min′, where v represents the original value of *sentiment*; min and max denote the minimum and maximum value of *sentiment* in the original dataset, which is -1 and 1, respectively; max′ and min′ denote the minimum and maximum value of rescaled *sentiment*, which is 1 and 5, respectively. For example, if the sentiment score v = 0, then the rescaled sentiment score on a 1 to 5 scale is 4 × [(v + 1)/2] + 1 = 2v + 3 = 3. This rescaling approach is in line with Hu. et al. [19].

Review length. Review length was operationalized as the number of words in a review.

Control variables. This study also controlled for several variables that could affect review helpfulness: *review age, review picture, is_Elite, profile photo, friend count,* and *review count. Review age* was measured by days elapsed after the review being posted. *Review picture* was operationalized as a dummy variable with 1 denotes the review text contains embedded picture(s) and 0 otherwise. *Is_Elite* was a dummy variable with 1 denotes that the reviewer was an Elite member and 0 otherwise. *Profile photo* was operationalized as a dummy variable with 1 represents the reviewer provided profile photo and 0 otherwise. *Friend count* was measured by the number of friends the reviewer has in Yelp, and *review count* was measured by the number of reviews the reviewer posted in the past. We also accounted for restaurant heterogeneity. This was done by including 99 dummy variables into our model to control for restaurant fixed effects, following the procedure of Chen and Lurie [23].

3.3 Model Specification

Since our dependent variable is a count measure, standard multiple regression is inappropriate. In addition, the variance of the dependent variable significantly exceeds its mean (mean = 1.30, variance = 2.99), suggesting a possible overdispersion problem. Then, the negative binomial model, which can account for overdispersion, is applied here. In line with prior research on review helpfulness (e.g., [23]), we employ the negative binomial regression to estimate the following model.

$$
\begin{aligned}
Helpfulness = \exp[&\beta_0 + \beta_1(rating) + \beta_2(sentiment) + \beta_3(length) \\
&+ \beta_4(rating \times sentiment) + \beta_5(rating \times length) \\
&+ \beta_6(sentiment \times length) + \beta_7(rating \times sentiment \times length) \\
&+ \beta_8(review\,age) + \beta_9(review\,picture) + \beta_{10}(profile\,photo) \\
&+ \beta_{11}(is\_Elite) + \beta_{12}(friend\,count) + \beta_{13}(review\,count) \\
&+ \rho + \varepsilon]
\end{aligned}
$$

In the above equation, ρ denotes the restaurant fixed effects, and ε is the idiosyncratic error.

4   Results

Table 1 reports the descriptive statistics and correlations for the main variables. The results of the negative binomial regression analyses are presented in Table 2. To alleviate the potential for multicollinearity, predictor and moderator variables (except those dummy variables) were mean-centered before creating the interaction terms, following the recommendation of Aiken and West [26]. Additionally, we examined the variance inflation factors for each model, and all were well below the generally acceptable threshold of 10 [27], suggesting that multicollinearity is not a problem for our analysis.

Table 1.   Descriptive statistics and variable correlations

Variables	Mean	SD	1	2	3	4	5	6	7	8	9
1. *helpfulness*	1.30	2.99									
2. *rating*	4.34	0.92	−.04								
3. *sentiment*	3.11	0.13	−.09	.16							
4. Ln (*length*)	4.24	1.08	.16	−.10	−.30						
5. Ln (*review age*)	6.38	1.10	.08	.02	−.06	.09					
6. *review picture*	0.08	0.27	.16	.04	−.04	.08	−.07				
7. *profile photo*	0.84	0.37	.13	.04	−.06	.08	.11	.10			
8. *is_Elite*	0.19	0.40	.21	−.01	−.07	.15	−.05	.18	.22		
9. Ln (*friend count*)	2.83	1.88	.33	.03	−.11	.17	.17	.22	.46	.50	
10. Ln (*review count*)	3.90	1.50	.27	−.03	−.12	.18	.20	.18	.35	.56	.71

In Table 2, Model 1 shows the baseline specification including the control variables. Model 2 displays the main effect, indicating that review rating (β = -0.185, $p < 0.001$) and text sentiment (β = −0.784, $p < 0.001$) are negatively and significantly related to review helpfulness, which respectively supports H1 and H2. Model 3 enters the two-way rating-sentiment interaction and shows that the interaction effect between review rating and text sentiment is positively significant (β = 0.437, $p < 0.001$). To facilitate interpretation, we plotted the interaction at positive (one standard deviation above the mean) and negative (one standard deviation below the mean) levels of review rating and text sentiment, following the procedure recommended by Cohen et al. [28]. As shown in Fig. 3, review rating has a negative effect on review helpfulness when text sentiment is negative, but this negative effect is weakened when text sentiment is positive; reviews with consistent rating-text valences are more helpful than reviews with inconsistent rating-text valences. Hence, H3 is supported.

Table 2. Results of negative binominal regression analyses

Variables	Model 1	Model 2	Model 3	Model 4
rating		-0.185***	-0.161***	-0.098***
		(0.006)	(0.006)	(0.007)
sentiment		-0.784***	-0.800***	-1.962***
		(0.053)	(0.053)	(0.083)
Ln (length)		0.162***	0.164***	0.138***
		(0.005)	(0.005)	(0.005)
rating × sentiment			0.437***	1.330***
			(0.054)	(0.079)
rating × Ln (length)				0.054***
				(0.005)
sentiment × Ln (length)				-0.628***
				(0.035)
rating × sentiment × Ln (length)				0.455***
				(0.032)
Ln (review age)	0.175***	0.173***	0.173***	0.173***
	(0.006)	(0.006)	(0.006)	(0.006)
review picture	0.344***	0.345***	0.344***	0.343***
	(0.019)	(0.018)	(0.018)	(0.018)
profile photo	0.252***	0.273***	0.278***	0.281***
	(0.021)	(0.020)	(0.020)	(0.020)
is_Elite	0.129***	0.100***	0.100***	0.100***
	(0.016)	(0.016)	(0.016)	(0.016)
Ln (friend count)	0.297***	0.305***	0.305***	0.303***
	(0.005)	(0.005)	(0.005)	(0.005)
Ln (review count)	0.046***	0.028***	0.030***	0.031***
	(0.006)	(0.006)	(0.006)	(0.006)
Number of Obs.	70,610	70,610	70,610	70,610
Log likelihood	-98327.913	-96840.938	-96806.073	-96513.398
Pseudo R^2	0.09	0.105	0.106	0.109

Note: Standard errors reported in parentheses. Restaurant dummy variables were included, but not reported for simplicity of presentation.* $p < 0.5$; ** $p < 0.01$; *** $p < 0.001$.

Model 4 in Table 2 further includes the three-way interaction of review rating, text sentiment, and review length. The significance of this three-way interaction ($\beta = 0.455$, $p < 0.001$) indicates that the interaction effect of review rating and text sentiment is strengthened as review length increases. To further examine this relationship, we split the sample by the median level of review length (median = 79). The median split is more robust than mean split, which is subject to possible impact of outliers. The outputs from these two regressions are included in Table 3. Model 5 and 6 in Table 3 indicate both the interaction effects of review rating and text sentiment for short and long reviews are positively significant. These interaction effects are plotted in Fig. 4.

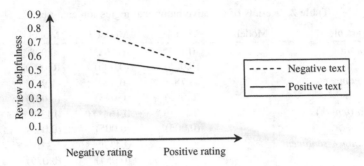

Fig. 3. Interaction of review rating and text sentiment on review helpfulness

Table 3. Results of rating-sentiment interaction for short reviews and long reviews

Variables	Model 5 Short reviews	Model 6 Long reviews
rating	-0.212*** (0.010)	-0.089*** (0.008)
sentiment	-0.617*** (0.056)	-2.729*** (0.146)
rating × *sentiment*	0.154*(0.055)	1.790*** (0.132)
Ln *(review age)*	0.211*** (0.010)	0.148*** (0.008)
review picture	0.413*** (0.033)	0.312*** (0.022)
profile photo	0.377 *** (0.031)	0.194*** (0.028)
is_Elite	0.191*** (0.028)	0.049** (0.019)
Ln *(friend count)*	0.274*** (0.007)	0.330*** (0.006)
Ln *(review count)*	0.057*** (0.009)	0.008 (0.008)
Number of Obs.	35,418	35,192
Log likelihood	-40285.486	-56096.459
Pseudo R^2	0.10	0.09

Note: Standard errors reported in parentheses. Restaurant dummy variables were included, but not reported for simplicity of presentation.* $p < 0.5$; ** $p < 0.01$; *** $p < 0.001$.

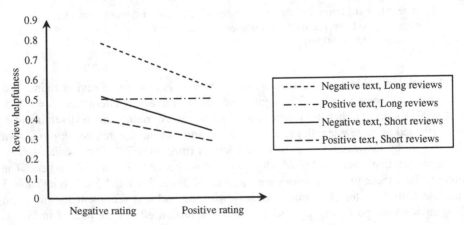

Fig. 4. Interaction of review rating, text sentiment and review lengh on review helpfulness

A comparison of short-review and long-review conditions shows that the interaction effect of review rating and text sentiment is more salient for long reviews than for short reviews, in support of H4.

5 Discussion and Conclusions

Review ratings and text sentiments respectively represent quantitative and qualitative aspects of user-generated product reviews. These two types of valence may not always be consistent, but complement each other in shaping consumers' review evaluation. The primary objective of this study is to understand and investigate the potential interaction effect between review rating and text sentiment on review helpfulness, and further specify whether this interaction effect is moderated by review length. With an analysis of 70,610 Yelp reviews across 100 restaurants, we find empirical evidence that both review rating and text sentiment have significant and negative effects on review helpfulness. Meanwhile, a positive interaction effect between review rating and text sentiment on review helpfulness is found. We also find that such interaction effect is stronger for long reviews than for short reviews.

This study has several theoretical contributions. First, this study adds to the WOM literature by documenting the negativity bias nature of text sentiment expressed in the reviews. This bias reported in previous studies on review helpfulness is mainly based on the impact of review rating [14, 23]. Few studies have provided empirical insights into the effect of text sentiment. We demonstrate that the well-accepted negativity bias effect applies to the valence of review text, such that reviews with negative text sentiments are perceived to be more helpful than those with positive text sentiments.

Second, we contribute to the literature by highlighting the interaction effect of review rating and text sentiment on review helpfulness. Previous studies mainly focus on identifying review and reviewer characteristics that may affect review helpfulness [1, 7–10, 12, 13]. Few extant studies investigate how review rating and text sentiment jointly influence consumers' perception of review helpfulness. Our study demonstrates a positive interaction between review rating and text sentiment. In addition, we shed light on the boundary conditions for the interaction of review rating and text sentiment by investigating whether this interaction effect is contingent on review length. We are the first to reveal the interaction of review rating and text sentiment on perceived review helpfulness is stronger for longer reviews.

Our findings also have several practical implications. First, our results indicate that the sentiments expressed in review texts do affect consumers' review helpfulness evaluation. In other words, consumers do rely on the textual contents in the reviews to make evaluations and decisions rather than rely solely on numerical features (such as review ratings). Firms may benefit by providing sentiment scores along with the numerical ratings [19], which not only facilitates consumers' review information processing, but also makes the inconsistency between review rating and text sentiment more salient and ultimately discounts negative information. Second, our findings suggest that consistent rating-text reviews are perceived to be more helpful than inconsistent ones. This highlights the importance for rating system designers to create functions that allow users to sort reviews based on the consistency between review ratings and text sentiments.

This study has several limitations that should be addressed in future research. First, as with many empirical studies in this domain (e.g., [23, 29]), this study has exclusively focused on a single category: restaurants. Future work would be needed to explore whether our findings would hold for other categories including, for example, books or movies. Second, we only consider one boundary condition (i.e., review length) for the interaction effect of review rating and text sentiment. It would be interesting for future work to identify other possible boundary conditions to provide more fine-grained understanding on review helpfulness.

References

1. Yin, D., Bond, S., Zhang, H.: Anxious or angry? Effects of discrete emotions on the perceived helpfulness of online reviews. MIS Q. **38**(2), 539–560 (2014)
2. Chevalier, J.A., Mayzlin, D.: The effect of word of mouth on sales: online book reviews. J. Mark. Res. **43**(3), 345–354 (2006)
3. Duan, W., Gu, B., Whinston, A.B.: The dynamics of online word-of-mouth and product sales—An empirical investigation of the movie industry. J. Retail. **84**(2), 233–242 (2008)
4. Forman, C., Ghose, A., Wiesenfeld, B.: Examining the relationship between reviews and sales: The role of reviewer identity disclosure in electronic markets. Inf. Syst. Res. **19**(3), 291–313 (2008)
5. Purnawirawan, N., De Pelsmacker, P., Dens, N.: Balance and sequence in online reviews: How perceived usefulness affects attitudes and intentions. J. Interact. Mark. **26**(4), 244–255 (2012)
6. Park, D.-H., Lee, J., Han, I.: The effect of on-line consumer reviews on consumer purchasing intention: The moderating role of involvement. Int. J. Electron. Commer. **11**(4), 125–148 (2007)
7. Cao, Q., Duan, W., Gan, Q.: Exploring determinants of voting for the "helpfulness" of online user reviews: A text mining approach. Decis. Support Syst. **50**(2), 511–521 (2011)
8. Mudambi, S.M., Schuff, D.: What makes a helpful online review? A study of customer reviews on Amazon. com. MIS Q. **34**(1), 185–200 (2010)
9. Schlosser, A.E.: Can including pros and cons increase the helpfulness and persuasiveness of online reviews? The interactive effects of ratings and arguments. J. Consum. Psychol. **21**(3), 226–239 (2011)
10. Schindler, R.M., Bickart, B.: Perceived helpfulness of online consumer reviews: The role of message content and style. J. Consum. Behav. **11**(3), 234–243 (2012)
11. Racherla, P., Friske, W.: Perceived 'usefulness' of online consumer reviews: An exploratory investigation across three services categories. Electron. Comme. Res. Appl. **11**(6), 548–559 (2012)
12. Korfiatis, N., García-Bariocanal, E., Sánchez-Alonso, S.: Evaluating content quality and helpfulness of online product reviews: The interplay of review helpfulness vs. review content. Electron. Commer. Res. Appl. **11**(3), 205–217 (2012)
13. Pan, Y., Zhang, J.Q.: Born unequal: A study of the helpfulness of user-generated product reviews. J. Retail. **87**(4), 598–612 (2011)
14. Sen, S., Lerman, D.: Why are you telling me this? An examination into negative consumer reviews on the web. J. Interact. Mark. **21**(4), 76–94 (2007)

15. Willemsen, L.M., Neijens, P.C., Bronner, F., de Ridder, J.A.: Highly recommended! the content characteristics and perceived usefulness of online consumer reviews. J. Comput.-Mediated Commun. **17**(1), 19–38 (2011)
16. Yang, J., Mai, E.S.: Experiential goods with network externalities effects: An empirical study of online rating system. J. Bus. Res. **63**(9), 1050–1057 (2010)
17. Ito, T.A., Larsen, J.T., Smith, N.K., Cacioppo, J.T.: Negative information weighs more heavily on the brain: The negativity bias in evaluative categorizations. J. Pers. Soc. Psychol. **75**(4), 887–900 (1998)
18. Ghose, A., Ipeirotis, P.G.: Estimating the helpfulness and economic impact of product reviews: Mining text and reviewer characteristics. IEEE Trans. Knowl. Data Eng. **23**(10), 1498–1512 (2011)
19. Hu, N., Koh, N.S., Reddy, S.K.: Ratings lead you to the product, reviews help you clinch it? The mediating role of online review sentiments on product sales. Decis. Support Syst. **57**, 42–53 (2014)
20. Tsang, A.S., Prendergast, G.: Is a "star" worth a thousand words? The interplay between product-review texts and rating valences. Eur. J. Mark. **43**(11/12), 1269–1280 (2009)
21. Bakhshi, S., Kanuparthy, P., Shamma, D.A.: Understanding online reviews: funny, cool or useful? In: Proceedings of the 18th ACM Conference on Computer Supported Cooperative Work & Social Computing, pp. 1270–1276. ACM, New York (2015)
22. Anderson, R.E.: Consumer dissatisfaction: The effect of disconfirmed expectancy on perceived product performance. J. Mark. Res. **10**(1), 38–44 (1973)
23. Chen, Z., Lurie, N.H.: Temporal contiguity and negativity bias in the impact of online word-of-mouth. J. Mark. Res. **50**(4), 463–476 (2013)
24. Pennebaker, J.W., Chung, C.K., Ireland, M., Gonzales, A., Booth, R.J.: The development and psychometric properties of LIWC2007. LIWC.net, Austin, TX (2007)
25. Ludwig, S., de Ruyter, K., Friedman, M., Brüggen, E.C., Wetzels, M., Pfann, G.: More than words: the influence of affective content and linguistic style matches in online reviews on conversion rates. J. Mark. **77**(1), 87–103 (2013)
26. Aiken, L.S., West, S.G.: Multiple regression: Testing and interpreting interactions. Sage Publications, Thousand Oaks (1991)
27. Hair, J.F., Anderson, R.E., Tatham, R.L., Black, W.C.: Multivariate data analysis, 5th edn. Prentice-Hall, Eenglewood, Cliffs, NJ (1998)
28. Cohen, J., Cohen, P., West, S.G., Aiken, L.S.: Applied multiple regression/correlation analysis for the behavioral sciences, 3rd edn. Lawrence Erlbaum Associates, New Jersey (2003)
29. Godes, D., Silva, J.C.: Sequential and temporal dynamics of online opinion. Mark. Sci. **31**(3), 448–473 (2012)

A Twitter View of the Brazilian Stock Exchange Market

Hugo S. Santos, Alberto H.F. Laender$^{(\boxtimes)}$, and Adriano C.M. Pereira

Department of Computer Science,
Universidade Federal de Minas Gerais, Belo Horizonte 31270-901, Brazil
{hugo,laender,adrianoc}@dcc.ufmg.br

Abstract. In this paper, we present a view of the Brazilian stock exchange market based on a large characterization and analysis of Twitter data. In our analysis, we show that events and news about the stock market are capable of generating peaks of publications by Twitter users and that the frequency of posts follows the starting of the exchange trading day and maintains for about three hours after the stock market closing hour. Moreover, based on a survey conducted with a specific niche of Twitter users, we have been able to estimate that 0.5 % of those users have some knowledge of the Brazilian stock market and are mostly individual investors interested in publishing and consuming news about this market, having 45 % of them used Twitter as a source for investment decisions. Finally, we have observed that the total number of orders and the financial volume are positively correlated for 66 % of the stocks mentioned on Twitter, whereas the oscillation and maximum oscillation dimensions present no correlation.

1 Introduction

Accurately understanding the behavior and trends of stock markets is still an enormous challenge [1–3]. Many researches have addressed this challenge aiming to predict the behavior of stock prices [4]. However, with the development of new technologies, particularly social networks like Twitter[1] and Facebook[2], new research opportunities have emerged due to the large amount of data generated by these networks that, in many circumstances, reflects the behavior and opinions of their networks' users on distinct subjects like economy and financial markets.

In this context, Twitter has emerged as the most popular media to propagate recent events [5]. The limit of 140 characters, initially seen as a constraint, has led users to disseminate posts quickly and succinctly. For example, on June 24, 2014, an information posted with exclusivity on Twitter demonstrated the capacity of this service to influence the trading on BOVESPA, the main stock exchange in Brazil and the eighth largest in the world. In this case, a single tweet

---

[1] http://www.Twitter.com
[2] http://www.facebook.com

© Springer International Publishing Switzerland 2015
H. Stuckenschmidt and D. Jannach (Eds.): EC-Web 2015, LNBIP 239, pp. 112–123, 2015.
DOI: 10.1007/978-3-319-27729-5_9

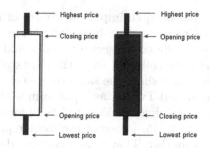

Fig. 1. Candlestick pattern.

posted by a State Governor reverted the upward trend of the stocks of COPEL, the electricity and power company of the state of Paraná[3].

In this paper, we present a Twitter view of the Brazilian stock exchange market based on a large scale characterization and analysis of almost three million tweets and more than 600 thousand official BOVESPA candlestick patterns for 73 stocks. These candlestick patterns describe the statistic oscillations of all stocks that compose IBOVESPA, the main BOVESPA index[4], at each 15 minutes (see Fig. 1). Thus, our main contributions in this paper are:

- A thorough static analysis based on Twitter data complemented by a survey conducted on a specific niche of its users. Recent studies in the literature characterize the information present on tweets and perform some kind of sentiment analysis, but do not assess whether the users who have posted them are in fact actual investors. We, on the order hand, have surveyed over 3,000 Twitter users in order to characterize them and know their specific interests, and found out that 45 % of them had used the social network as a source of information to make investment decisions.
- A detailed temporal and cross correlation analysis that characterizes the Twitter users' posting patterns and how the number of IBOVESPA related posts reflects the oscillation of the stocks in the market, specifically showing the stocks that can be monitored through the social network.

The remainder of this paper is organized as follows. Section 2 discusses related work. Section 3 describes the methodology underlying the analysis conducted in Sect. 4. Finally, Sect. 5 presents our conclusions and future work.

2   Related Work

The work presented in this paper relates to a broad spectrum of research on online social networks [6,7], stock markets [8–10] and, more specifically, Twitter [5,11,12].

---

[3] Infomoney: Paraná's governor vetoes through Twitter COPEL's price realignments; stocks turn down - http://goo.gl/T4YTEt

[4] IBOVESPA stocks are responsible for 80 % of the BOVESPA financial movement.

Particularly, Twitter has been an important source of real time information for several studies that aim to analyze behavior patterns and market trends. Sharma et al. [13] have developed a service to infer and characterize Twitter user attributes by using data collected from their lists of interest (a user can create a list of other users that share the same interests). Cheong et al. [14], on the other hand, have addressed Twitter as a platform with potential to provide collective intelligence by obtaining opinions and facts for decision making.

The work by Bollen et al. [15] is one of the most representative studies on the North American stock market. They have investigated whether the collective state of mood derived from the analysis of a large volume of Twitter data is correlated with changes in the value of the Dow Jones index (DJIA) over time. As a result, they have found an accuracy of 87.6 % in anticipation of changes in daily DJIA closing. Sprenger et al. [16] have studied Twitter as a platform to exchange information about the stock market. They have applied computational linguistic methods to analyze about 250,000 tweets related to the stock market. In their work, they have found associations between the volume of tweets and the volume of transactions as well as between the feelings expressed by the tweets and the stock returns, thus concluding that users that post more tweets on investment advice have on average more followers and receive more re-tweets.

Finally, the main differential of this work is an exclusive view of the Brazilian stock market that shows which BOVESPA stocks can be directly monitored through Twitter and a detailed characterization of those users who are in fact actual investors, an issue that has not been exploited before in the literature.

3   Methodology

In order to perform our analysis, we collected data from Twitter using its REST API[5]. We considered tweets posted between July 2013 and July 2014. The data collection was carried out in collaboration with InWeb - The National Institute of Science and Technology for the Web, a Brazilian research initiative that aims to foster the development of new technologies for the Web. For BOVESPA, we used data collected through its official Web service[6] by one of our research partners. We used a 15-minute candlestick for each stock that forms IBOVESPA, restricting this dataset to the same period of the Twitter dataset.

For the purposes of our analysis, tweets have been classified as relevant when they included an explicit IBOVESPA stock code in its text. As an example, the tweet in Portuguese "*@clubedopairico PETR4 nao foi a R$13.65 em 21/11/2008?*" is part of our dataset because it explicitly mentions the code *PETR*4, which is the IBOVESPA id for PETROBRAS, the Brazilian oil company. Table 1 shows basic statistics of our Twitter and BOVESPA datasets.

In the next section, we present our experimental analysis based on the Twitter and BOVESPA datasets.

---

[5] https://dev.twitter.com/docs/api
[6] http://goo.gl/1SWujW

Table 1. Main statistics of our datasets.

Twitter Dataset	2,973,410 tweets
First collected tweet	07/01/2013 00:22:03
Last collected tweet	07/31/2014 23:10:24
BOVESPA Dataset	619,125 candlesticks
First collected candlestick	07/01/2013 10:00:00
Last collected candlestick	07/31/2014 17:45:00

4 Experimental Analysis

In this section, we present a thorough analysis of the collected tweets, which aims to answer the following research questions:

Q1: What is the volume of data shared on Twitter about the Brazilian stock market? Is there a specific niche among the Twitter users? If yes, are they actual investors?

Q2: Which temporal and behavioral patterns are reflected on the Twitter dataset over time?

Q3: Is there a correlation between the oscillation of the stocks and their respective number of mentions in Twitter posts?

To answer these research questions, we have divided our analysis in three parts. For the first research question (Q1), static analysis (Sect. 4.1), we present only general statistics of the tweets and the survey results, disregarding any temporal aspect. For the second research question (Q2), temporal analysis (Sect. 4.2), we complement the static analysis by considering behavioral patterns over the 13 months of data. For the third research question (Q3), correlation analysis (Sect. 4.3), we consolidate our analysis by evaluating the correlation between Twitter posts and BOVESPA data, considering the number of posts that contain any mention to a stock and the variation of each one of the following four dimensions: (a) the volume of orders (number of orders sent by all investors), (b) the daily oscillation (Fig. 1: $ClosingPrice - OpeningPrice$), (c) the maximum daily oscillation (Fig. 1: $HighestPrice - LowestPrice$) and (d) the total daily financial movement (number of stocks multiplied by their value at the order creation time).

4.1 Static Analysis

In this section, we address research question Q1 by analyzing the tweet attributes that explicitly mention some IBOVESPA stock code.

Table 2 shows the main characteristics of the dataset. As we can see, only 3.1 % of the tweets can be classified as related to the stock market, i.e., they explicitly mention a stock code. Moreover, these tweets have been posted by a

Table 2. Static analysis of tweets.

#tweets that mention a stock	93,179 (3.1 % of 2,973,410)
#stock-interested users	4,816 (0.5 % of 1,194,771)
#retweets	4,467 (5.00 %)
#replies	2,657 (2.85 %)
#tweets containing a URL	40,293 (80.00 %)

Table 3. Survey form questions

1) How do you fit yourself in relation to the Brazilian stock market?

a) I am investor	67%
b) I am not an investor, but I intend to invest in the future	28%
c) I am an investor but do not intend to invest in the future anymore	5%

2) Which one of the following options best describes you?

a) Individual Investor	74.8%
b) Independent Investment Agent	6.3%
c) Professional of Specialized Media	4.2%
d) Financial Analyst	3.5%
e) Professional of Investors Relationship	0.7%
f) None of above	10.5%

3) In the stock market context, do you post tweets more frequently about what?

a) Your personal opinion:	30,8%
b) Technical analysis:	9.1%
c) News:	39.9%
d) Other subjects:	3.5%
e) I don't post:	16.8%

4) In general, what kind of post from other users you consider as more relevant?

a) Personal opinions	15,4%
b) News	30,1%
c) Finantial analysis	49,7%
d) Other subjects	2.8%
e) I don't use Twitter	16.8%

5) Are the tweets that you post about the stock market in general related to your own portfolio?

a) Yes	27.3%
b) No	51.0%
c) I don't post tweets about the stock market	21.7%

6) Have you used or use information from Twitter to decide about some investment?

a) Yes:	44.8%
b) No:	55.2%

niche of 0.5 % of the Twitter users considered, which means that this specific niche of users is actually interested in the stock market.

According to metrics adopted by Benvenuto et al. [17], our figures for number of retweets and replies, and the large number of links found in these tweets suggest a behavioral pattern characterized by posts from users who are primarily interested in disclosing new information (news). Based on these findings, we have decided to conduct a survey to better know who are these users.

Thus, we have extended our static analysis in order to investigate whether users who post messages related to the financial market are potential investors and what are their main interests. For this, we selected 3,302 users from those

4,816 interested in stocks who have posted at least once in the period from January, 1 to July, 31 2014, and tweeted them the link of a short Google survey form in order to characterize their behavior in the social network. The questions (1 to 6) and the compilation of the responses returned by 225 users are shown in Table 3.

From the survey it was possible to see that 67 % of the users are investors (Question 1) and that most of them actually judge themselves as individual investors (Question 2: Individual investor - 74.8 %). We can also see that these users tweet more frequently about news and their personal opinions (Question 3) and that they are mainly interested in receiving news and financial analysis (Question 4). Moreover, 44.8 % of them use the social network data to support decisions about their own investments (Question 6). Finally, only 27.3 % of the users mention their own stocks when tweeting about the market, which means that, despite such mentions, most users might not own any stocks (Question 5).

Among other insights obtained from the survey, we have verified that the ability of Twitter to influence and be influenced by the behavior of the BOVESPA stocks is due to the news shared among the users. Considering that the Efficient Market Hypothesis (EMH) [18] suggests that stock prices are more influenced by new information than by present and past prices, we can say that there is a behavioral pattern among users who use Twitter as an information channel that might influence the financial markets and vice versa.

4.2 Temporal Analysis

In the previous section, we performed a static analysis of our Twitter dataset based only on the tweet attributes. In this section, we address research question Q2 by analyzing this dataset throughout the period of 13 months. More specifically, we analyze how the daily posts vary along the weekdays and the BOVESPA trading days, the number of posts by hours of the day and, finally, the segmentation of engaged users responsible for the posts.

The frequency of posts may vary according to the users' daily routines, trading days and weekdays. From Fig. 2, we clearly notice how the pattern of posts follows the BOVESPA trading days, having a regular number of posts during the weekdays and almost a irrelevant volume during weekends and holidays.

Looking at Fig. 2, we also clearly identify peaks of Twitter posts associated with significant events in the stock market. For example, the peak on July 31, 2014 can be explained by a sequence of positive results published by Brazilian companies in the second quarter. These results include a 280 % profit of Vale Corporation (VALE5), a reverse of loss into a significant profit by EMBRAER (EMBR3), and an increase of 16 % of Ambev's (ABEV3) profit due to the World Cup held in Brazil. In addition, the peak of tweets on October 31, 2013 is related to the withdrawal of OGX stocks from the IBOVESPA[7].

---

[7] Economic Value - http://goo.gl/6UVav9

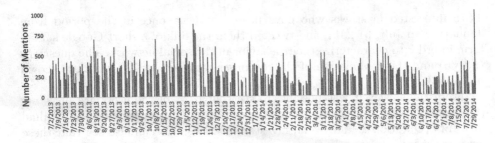

Fig. 2. Number of tweets per day over 13 months.

To better understand the patterns for user posts, we have also analyzed the distribution of tweets in a smaller granularity of time. Thus, Fig. 3 shows the distribution of tweets through the day hours over our period of analysis.

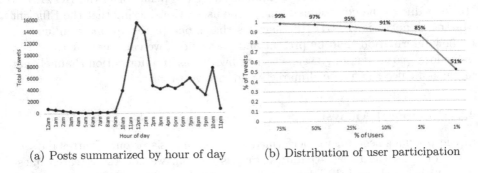

(a) Posts summarized by hour of day   (b) Distribution of user participation

Fig. 3. Posts summarization and distribution.

It is known that the market starts its trading at 10am and ends it at 5pm. Thus, from Fig. 3(a) we observe that the most engaged period for users posting on Twitter is between 11am and 13pm. In addition, the posting remains after the market trading closing due to factors such as external markets, related news published and, mainly, strategic information published by companies that occur after the market closes, according to CVM regulations[8].

Now we focus our analysis on the distribution of user participation. For this, we have plotted the distribution of posts for our entire period of analysis. From the cumulative distribution in Fig. 3(b), we can observe a pattern that is typical in many of Web systems. In this case, 10 % of our users are responsible for generating more than 90 % of the collected tweets. It suggests that there is some groups of users with high engagement in content generation. We believe that these users act like *influencers* in the social network by posting information to the other 90 % of users.

---

[8] Monetary and Exchange Commission - http://www.cvm.gov.br

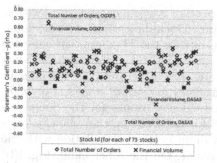 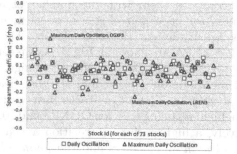

(a) Rate for number of orders and financial volume dimensions.

(b) Rate for daily oscillation and maximum daily oscillation dimensions.

Fig. 4. Spearman coefficient calculated for each stock.

Table 4. Cohen's scale.

Correlation Coefficient	Description
0.0-0.1	trivial, very small, insubstantial
0.1-0.3	small, low, minor
0.3-0.5	moderate, medium
0.5-0.7	large, high, major
0.7-0.9	very large, very high, huge
0.9-1.0	nearly, practically, or almost: perfect, distinct, infinite

Table 5. Our positive-negative Cohen's based scale.

Sperman Coefficient	Description
$-1 \leq \rho \leq -0.1$	negatively correlated
$-0.1 \leq \rho \leq 0.1$	uncorrelated
$0.1 \leq \rho \leq 1$	positively correlated

4.3 Correlation Analysis

In the last two sections, we focused our analysis only on the Twitter data. In this section, we address research question Q3 by performing a correlation analysis between the daily behavior of the BOVESPA stocks and the frequency of mentions to these stocks on Twitter. Our analysis is based on the Spearman correlation coefficient [19] and the Cohen's scale [20].

The Spearman correlation coefficient between two variables X and Y considers a continuous scale from -1 to 1, where -1 and 1 are boundary values on which the two variables can be interpreted as perfectly correlated. Table 4 presents the Cohen's scale. Based on these two concepts, we define the correlation range shown in Table 5 for the purpose of identifying a positive, negative or nonexistent correlation, without, necessarily, considering the magnitude of that correlation.

For our correlation analysis, we have built a time series for each of the four stock dimensions, i.e., total number of orders, financial volume, daily oscillation (percentage of price variation) and maximum daily oscillation, in order to compare them with the number of respective stock mentions on Twitter. Then, for each pair of series "tweet-mentions, stock-dimension" we calculated the Spearman correlation coefficient. The results for the 73 stocks that compose IBOVESPA are shown in Fig. 4. As we can see in Fig. 4(a), for some stocks, such as OGXP3 and DASA3, there is a significant correlation between the number of mentions on Twitter and the dimensions *total number of orders* and *financial volume*. These two stocks have reached extreme values for the Spearman coefficient, being 0.66 for OGXP3 and -0.36 for DASA3, i.e., they show, respectively, a high positive correlation and a medium negative correlation with the *total number of orders* and the *financial volume*. On the other hand, Fig. 4(b) shows extreme values of approximately 0.40 for OGXP3 and -0.25 for LREN3, indicating that the number of mentions of these two stocks on Twitter have, respectively, a moderate positive correlation and a low negative correlation with the dimension *maximum daily oscillation*.

However, results in Fig. 4(a) and (b) show that most of the stocks present "uncorrelated" values for the Spearman coefficient and that these values are divergent depending on the stock and dimension considered. Thus, in order to deepen our analysis, we have calculated for each dimension its total correlation percentage with respect to the total number of stock mentions through out the Twitter daily mention series limited by our Positive-Negative scale. To do so, we first considered 100 % of the days analyzed and then selected only those stocks including some mention at least every other day. The results are shown in Tables 6 and 7.

Table 6. Percentage of correlation for each dimension considering the set of all tweets including a stock mention.

Dimension (100 %)	Correlation of 73 stocks		
	Negative	Uncorrelated	Positive
Total Nr. Orders (#orders)	5 %	29 %	66 %
Finan. Volume ($)	1 %	25 %	74 %
Daily Osci. (%)	5 %	66 %	29 %
Max. Daily Osci. (%)	10 %	58 %	33 %

Table 7. Total correlation of each dimension for the set of tweets with mentions.

Dimension (50 %)	Correlation of the 31 filtered stocks		
	Negative	Uncorrelated	Positive
Total Nr. Orders (#orders)	3 %	26 %	71 %
Finan. Volume ($)	0 %	16 %	84 %
Daily Osci. (%)	3 %	58 %	39 %
Max. Daily Osci. (%)	6 %	48 %	45 %

The results presented in Tables 6 and 7 show that there is a positive correlation between the total number of orders and the financial volume with respect to Twitter stock mentions. That is, as the volume of financial trading increases the number of stock mentions in tweets also grows and vice versa. On the other hand, both tables show that there is no correlation between the daily and maximum daily oscillation dimensions with the daily total number of stock mentions in tweets and vice versa.

These results suggest that, for dimensions financial volume and total number of orders, Twitter users may get some indicators from the social network about the behavior of BOVESPA stocks. It is noteworthy that these two dimensions are important for the financial market because they might be used by investors as indicators of stock financial liquidity.

Table 8. The set of stocks that are positively correlated with financial volume and total number of orders.

Positive	EMBR3, OGXP3, BBDC4, VIVT4, BVMF3, BBDC3, LLXL3, CRUZ3, LIGT3, BTOW3, GFSA3, USIM5, ITUB4, SBSP3, OIBR4, ALLL3, BBAS3, VALE5, PETR3, PETR4, CSNA3, PDGR3
Uncorrelated	MMXM3, BISA3, MRVE3, NATU3, BRML3

Table 8 shows the set of stocks that are correlated with both financial volume and total number of orders. This table includes the 22 stocks with positive correlation, which were considered relevant by our group of Twitter users. It includes, for example, stocks of oil and steel companies such as Petrobras (PETR4), Usiminas (USIM5) and Vale (VALE5), as well as major banks such as Bradesco (BBDC4), Banco do Brasil (BBAS3) and Itau Unibanco (ITUB4). We have also found that the stocks identified by MMXM3, BISA3, MRVE3, NATU3 and BRML3, in general, do not contain a pattern of posts correlated with anyone of the dimensions considered, i.e., there is no evidence to infer any pattern of Twitter posts that would be useful as an indicator for investors interested in these five stocks. Finally, the last four stocks are neither negative nor positively correlated with both dimensions at the same time. In this case, a negative correlation cannot be considered conclusive because we have not analyzed absolute values for the time series.

As a final comment, from our results in this section, it is noteworthy that they also demonstrate that the most popular stocks (those known as *bluechips*) are the ones most mentioned on Twitter and, therefore, tend to show some correlation with their behavior on BOVESPA.

5   Conclusions

In this paper, we have presented an analysis of the Brazilian stock exchange market based on Twitter data collected from July, 2013 to July 2014. This analysis comprises an initial characterization of static patterns revealed by the attributes of the users' posts, a temporal analysis of the daily posts along a periodic of 13 months, and a correlation analysis between the daily behavior of the BOVESPA stocks and the frequency of mentions to these stocks on Twitter.

From the static analysis we have found that only 3.1% of the tweets posted during the considered period were classified as related to the stock market, i.e., these tweets explicitly mentioned a BOVESPA stock code in its text. Furthermore, we have also found a specific niche of 0.5% of users that are actually interested in the stock market, of which 45% of them actually use Twitter as an information source for investment decisions.

Our temporal analysis has showed patterns of user behavior and frequency of tweets. As a major finding, our analysis has shown that events and news about the stock market are capable of generating publication peaks on Twitter. In addition, we have found out that the frequency of publications on Twitter are mainly concentrated on the business hours of the exchange trading days, but the frequency of posts is maintained for about five hours after the stock market closing hour. Moreover, we have also found out that 10% of the users are responsible for more than 90% of the Twitter posts.

Finally, in the correlation analysis we have studied the dependence between the daily behavior of the BOVESPA stocks and the frequency of stock mentions on Twitter. Initially, we calculated the Spearman correlation coefficient for the IBOVESPA's stocks, considering time series involving the number of daily stock mentions on Twitter and each one of the four dimensions: total number of orders, financial volume, daily oscillation and maximum daily oscillation. In this analysis, we have observed that the total number of orders and the financial volume are positively correlated with 66% of the stocks mentioned on Twitter, whereas the daily oscillation and maximum daily oscillation dimensions show no correlation. As a final result, we have found 22 stocks with positive correlation with the number of daily stock mentions, which can be considered as the most relevant ones to our group of Twitter users.

The stock market and Twitter patterns revealed by our static, temporal and correlation analyses have a broad applicability as investment indicators. Thus, as future work we plan to exploit a graph model to explain the behavior of our set of Twitter users based on concepts such as centrality, clustering and communities. In addition, we intend to expand this work by carrying out a sentiment analysis on our Twitter dataset in order to investigate how the users' opinion correlates with the stock oscillations in the Brazilian market.

Acknowledgments. This research was supported by the Brazilian National Institute of Science and Technology for the Web (InWeb - CNPq grant number 573871/2008-6) and by individual grants from CNPq and Fapemig.

References

1. Atsalakis, G.S., Valavanis, K.P.: Surveying stock market forecasting techniques - Part II: Soft computing methods. Expert Syst. Appl. **36**(3), 5932–5941 (2009)

2. Martinez, L.C., da Hora, D.N., de M. Palotti, J.R., Meira Jr, W., Pappa, G.L.: From an artificial neural network to a stock market day-trading system: a case study on the BM&F BOVESPA. In: Proceedings of the International Joint Conference on Neural Networks, pp. 2006–2013 (2009)
3. Silva, E., Castilho, D., Pereira, A., Brando, H.: A neural network based approach to support the market making strategies in high-frequency trading. In: Proceedings of the International Joint Conference on Neural Networks, pp. 845–852 (2014)
4. Fama, E.F., et al.: The adjustment of stock prices to new information. Int. Econ. Rev. **10**(1), 1–21 (1969)
5. Kwak, H., Lee, C., Park, H., Moon, S.: What is Twitter, a social network or a news media? In: Proceedings of the 19th International Conference on World Wide Web, pp. 591–600 (2010)
6. Maia, M., Almeida, J., Almeida, V.: Identifying user behavior in online social networks. In: Proceedings of the 1st Workshop on Social Network Systems, pp. 1–6 (2008)
7. Mislove, A., Marcon, M., Gummadi, K.P., Druschel, P., Bhattacharjee, B.: Measurement and analysis of online social networks. In: Proceedings of the 7th ACM SIGCOMM Conference on Internet Measurement, pp. 29–42 (2007)
8. Zhang, W., Skiena, S.: Trading strategies to exploit blog and news sentiment. In: Proceedings of the Fourth International Conference on Weblogs and Social Media, pp. 375–378 (2010)
9. Ruiz, E.J., Hristidis, V., Castillo, C., Gionis, A., Jaimes, A.: Correlating financial time series with micro-blogging activity. In: Proceedings of the Fifth ACM International Conference on Web Search and Data Mining, pp. 513–522 (2012)
10. Gilbert, E., Karahalios, K.: Widespread worry and the stock market. In: Proceedings of the International Conference on Weblogs and Social Media, pp. 59–65 (2010)
11. Cha, M., Benevenuto, F., Haddadi, H., Gummadi, P.K.: The world of connections and information flow in Twitter. IEEE Trans. Syste., Man, Cybern., Part A **42**(4), 991–998 (2012)
12. Sprenger, T.O., Tumasjan, A., Sandner, P.G., Welpe, I.M.: Tweets and trades: the information content of stock microblogs. Eur. Financ. Manag. **20**, 926–957 (2013)
13. Sharma, N.K., Ghosh, S., Benevenuto, F., Ganguly, N., Gummadi, K.: Inferring who-is-who in the Twitter social network. ACM SIGCOMM Comput. Commun. Rev. **42**(4), 533–538 (2012)
14. Cheong, M., Lee, V.: Integrating web-based intelligence retrieval and decision-making from the twitter trends knowledge base. In: Proceedings of the 2nd ACM Workshop on Social Web Search and Mining, pp. 1–8 (2009)
15. Bollen, J., Mao, H., Zeng, X.: Twitter mood predicts the stock market. J. Comput. Sci. **2**(1), 1–8 (2011)
16. Sprenger, T.O.: Tweettrader.net: leveraging crowd wisdom in a stock microblogging forum. In: Proceedings of the International Conference on Weblogs and Social, The AAAI Press (2011)
17. Benevenuto, F., Magno, G., Rodrigues, T., Almeida, V.: Detecting spammers on Twitter. In: Proceedings of the 7th Collaboration, Electronic Messaging, Anti-Abuse and Spam Conference, July 2010
18. Fama, E.F.: The behavior of stock-market prices. J. Bus. **38**(1), 34–105 (1965)
19. Zar, J.H.: Significance testing of the spearman rank correlation coefficient. J. Am. Statis. Assoc. **67**(339), 578–580 (1972)
20. Cohen, J., Cohen, P., West, S.G., Aiken, L.S.: Applied Multiple Regression/Correlation Analysis for the Behavioral Sciences. Routledge, London (2013)

Process Management

Towards Smart Logistics Process Management

Raef Mousheimish[1,2]([✉]), Yehia Taher[1], and Béatrice Finance[1]

[1] Laboratoire PRiSM, Université de Versailles Saint-Quentin-en-Yvelines,
78000 Versailles, France
{raef.mousheimish,yehia.taher,beatrice.finance}@prism.uvsq.fr
[2] Fondation des Sciences du Patrimoine, LabEx PATRIMA, Cergy, France

Abstract. Logistics processes are generally agreed-upon, long running propositions between multiple partners, which are specified over Service Level Agreements as constraints to be maintained. However, these constraints can be violated at any time due to various unforeseen events that may stem from the process evolving context, leading the process to end up in unfortunate situations. In this paper, we present our framework that correlates critical business operations together with contextual events in order to predict possible violations prior to their occurrences while proactively generating mitigation countermeasures. In addition we develop a software and experiment it to demonstrate the practical applicability of the framework.

Keywords: Business process management · SLA violations · Prediction · Adaptation · SLA Violations

1 Introduction

Logistics is among the more vital economic activities for every business firm, and its management is deemed to be highly challenging. Cross-organizational business collaborations within a logistics process are usually stipulated in forms of mutual commitments that involved parties accept to integrate *(e.g., delivery time≤ 2 days)*. In any business, customer satisfaction comes first, thus ensuring compliance to the agreed-upon SLA stipulations such as *delivery time*, is a key concern. Therefor parties involved in a logistics process are always aiming at rendering their services as speedily as possible to enforce their reputation and competitiveness in global trades. However, this is a non-trivial goal, as in real-world situations, the conditions under which a logistics process is running (e.g., location, road status) may evolve stochastically over time. Falling short to efficiently handle these dynamic changes, can lead the process to unfortunate situations. In particular, a remarkable amount of logistics operations embody either late deliveries, or late carriers cancellations [4].

In line with the Internet of Things *(IoT)* evolution, which promotes today an increasing availability of event data from mobile and sensor devices, and seamless connection to anyone, anywhere, a range of research proposals have explored the

© Springer International Publishing Switzerland 2015
H. Stuckenschmidt and D. Jannach (Eds.): EC-Web 2015, LNBIP 239, pp. 127–138, 2015.
DOI: 10.1007/978-3-319-27729-5_10

problem of SLA monitoring from contextual perspectives. However, these monitoring approaches are mostly reactive as they alert a violation after it occurs. Only few approaches highlight the importance of predictive monitoring where a violation is identified prior in time and a proactive adaptation is engaged to avoid the violation [2,5,7]. Generally, these approaches try to efficiently implement the concept of the MAPE *(Monitor-Analyze-Plan-Execute)* loop, introduced by Kephart and Chess in their vision on autonomic computing [1]. The loop is considered as a cornerstone to achieve proactive computations, as it drew theoretically-feasible solutions. However when it comes to its realization, research approaches are still suffering from numbered limitations, where they are still reactive rather than proactive, and they lack dynamic adaptation policies (each exception needs to be addressed at design-time).

In this paper, we present a new framework that aims at efficiently predicting possible time violations, and proactively address them by dynamically generating adaptation countermeasures. To achieve reliable predictions, we specified a prediction algorithm which is mainly constructed upon a delay propagation mechanism using monitored events of interest. The algorithm makes it possible to predict the status of the process while pointing out any possible violations before they occur. To proactively mitigate predicted violations, we propose a new algorithm, called the sub-goals algorithm, in charge of looking for available means to avoid such violations. The algorithm works on dividing the process into fragments according to their scopes in order to identify the smallest process fragment to be adapted. Such a technique allows to keep the adaptation as limited as possible. Pointing out violations from a predicted state of the process (not the current one), and using local adaptation techniques are novel mechanisms that could be propagated to more fields to better handle business process managements. The framework has been implemented and tested using a logistics use case.

The plan of the paper is the following. In Sect. 2, we motivate our work by introducing a logistics scenario and reviewing the shortcomings in the state of the art. In Sect. 3, we discuss our framework. In Sect. 4, we sketch out our demonstration prototype and experiments. Finally, we conclude and envision future researches in Sect. 5.

2   Motivations

First we outline a motivating scenario, in the domain of multi-modal transportation, which we will use as a running example throughout the paper. Afterward, we review related works and highlight their shortcomings.

Fig. 1. The initial plan

2.1   A Logistics Scenario

Let's assume a client in Valencia, Spain, requests some time-sensitive shipment from a manufacturer in Maastricht, the Netherlands. We assume that the freight will be ready for transport at 8:00 am, and needs to be available for the client before 8:30 pm, or else, penalties need to be applied. A specialist came with the least expensive plan as depicted in Fig. 1. Let's consider that during the execution of the first task, an unexpected traffic jam on the road delayed the shipment arrival to Brussels airport by one hour. To add insult to injury, we assume that high security checks at Brussels airport also added half an hour of delay and thus, the shipment couldn't make it to the flight on time. To deal with such a situation, one possible solution may consist in waiting for the next flight to Madrid. The freight will end up by arriving there but with heavily delays.

Although this seems to be a pretty reasonable solution, it clearly shows a considerable limitation and an inability to properly serve the client request, i.e., to comply to the agreed-upon conclusions and deliver the freight on time.

Failing to proactively adapt the process to efficiently handle unforeseen contextual events, the scenario shows how unfortunate the situations the process may end up in. The scenario highlights the need of a next generation of logistics processes, i.e., smart processes that are able to monitor their own performance by sensing and interacting with the physical word.

2.2   Related Work

In the PREvent framework [3] authors define a set of prediction checkpoints over the flow of the process. Each checkpoint is associated with a predictor and an adaptor. Every time one of these points is reached, the proper predictor is triggered to check for probable future violations, and the adaptor then executes one of the statically predefined adaptation actions if needed. The monitoring and predictions are triggered only upon checkpoints and the adaptor component is limited to a set of predefined actions that need to be designed before execution. Both the definition of good checkpoints and the assignment of good adaptation actions at design-time, are tedious and human-oriented tasks that are susceptible to erroneous configurations.

Authors in [6] introduced what is called *planlets* processes, where tasks are annotated with pre-conditions, desired effects, and post-conditions. The pre- and post-conditions are checked respectively before and after a task execution to catch violations and so violations can not be anticipated neither detected during the execution.

The approaches described above are general approaches in the domain of business process management, we look now at more specific projects in the area of transportation.

In [4], authors addressed the problem of delays in logistics use-cases, however they mainly rely on the absence of events declaring that the process reached a specific milestone. In our work we introduce another mechanism that could predict delays in a faster way, leaving more time to adapt.

The Green European Transportation Service or the GET Service[1], is one of the most recent European projects held in the domain of logistics. It incorporates efforts to enhance business processes in order to support more efficiently transportation missions. In particular, they propose predictive monitoring in order to detect violations prior to their occurrences in an online logistics process [8]. The focus is mainly on flight shipping activities, where they defined constrained (flight position) and monitored (altitude and speed) attributes. During execution, they monitor these attributes in an interval-based scheme. By counting on a trained support vector machine classifier, they detect divergence if any. Although this approach is partially related, yet it is very specific to detecting in advance if a flight is going to land in another airport than planned. As shown in our motivating scenario we are more interested in a multi-modal transportation process.

Finally, authors in [9] modeled a logistics process for handling client requests in the domain of multi-transfer container transportation. However the approach only handled client request events, but not unexpected events that may arise during the execution. Second, they counted on a predefined repository that contains statically-calibrated transportation network information, e.g., the time to go from point A to B is constant. This seems to be unrealistic as very often, there are delays as mentioned in the study conducted in [4].

Shortcomings Analysis: Many research initiatives usually count on the availability of predefined adaptation actions. They let the end user defines a static compensation plan at design-time for every unexpected situation that may occur. However we argue that it is not possible to enumerate all possible situations that may happen especially in a dynamic and evolving environment like logistics. Thus, these initiatives fall short in enabling the process to handle unforeseen situations.

Another point to highlight is that current approaches only protect the global goal by monitoring the deviation of local SLAs or constraints. They are mainly detecting and not predicting local violations. Like employing checkpoints, computing the response time of services after their execution, or annotating tasks with pre- and post-conditions, whereas continuous monitoring would be more suitable and required.

3   Smart Logistics Processes

In this section we present the details of the proposed framework as sketched in Fig. 2. The *Plan* component is an external module that could generate a logistics process. The *Execute* component is in charge of tracking the execution. The ongoing process is then monitored with respect to some identified events, called the events of interest, by the *Monitor* component. Then, the *Predict* component is in charge of drawing anticipations about the future state of the process. In case any future violation is predicted, the *Recommend* component is then triggered

---

[1] http://getservice-project.eu/

and an alternative solution is dynamically generated and recommended to the user. The user can approve or reject the solution. In case of rejection, the system will loop again to the *Recommend* component and look for another solution. Otherwise, the system moves into the *Adapt* phase where the initial process is proactively adapted at run-time to avoid the predicted violation. As depicted, this module can query the planner again if needed. In the following, we focus on the *Predict, Recommend,* and *Adapt* components to answer the question: how to predict violations and proactively mitigate them?

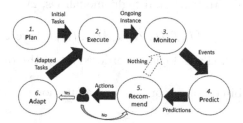

Fig. 2. The overall framework

3.1   Preliminary Definitions

In logistics, workflow is widely adopted to handle the flow of documents. However it is important to particularize that in this work, we are interested in managing the *flow of work*. That is, we are not interested in signing, checking, import/export, etc. The focus is on the real shipping actions, *the monitorable tasks*, that requires continuous monitoring as described in [8]. Taking this into account, in the domain of multi-modal transportation operations, the processing of client requests can be done by generating a sequence of specific activities [9]. In the rest of this paper we will be focusing on three essential tasks as described below, which are sufficient to fulfill requests. The **load** (L) and **unload** (U) tasks, they represents the (un)loading of the shipment (For flights, we assume that airport security checks are incorporated within these tasks for brevity purposes). The **transport** task (T), it is about transporting the freight from one place to another.

The Annotated Time-Aware Representation: For a better analysis of logistics processes, we define an annotated time-aware representation (Gantt-like charts). It simplifies the extraction of any needed information regarding each task (e.g., completion time, permitted delay time, location). Figure 3(a) illustrates this representation by putting our scenario into practice. The empty box, shows an extra unallocated time. The black boxes represent the *Permitted Delay Time* (PDT) of each task. These delays show how much a task *End Time* (ET) can exceed its agreed-on value. They depend essentially on the next task in the flow and the *Global Completion Time* (GCT). For now, these PDTs can be considered as a set of configurations that could be manipulated by users. Using this time-aware representation, each task could be projected into the time domain to extract its *Start Time* (*ST*), *End Time* (*ET*) and also its *Permitted Delay Time* (*PDT*). Note that in other works such as [9], our *PDT*s look like earliest and latest times. However our proposition differs by specifying a *PDT* for each task, not just for the start and end of the global process. This constitutes the main enabler to detect even local violations, as it will pave a way for more flexibility as shown later.

Formal Notations: The Δ symbol stands for the domain of all tasks within a logistics process (again to handle the flow of work). For now, as we are supporting three tasks, in addition to truck and flight modalities, Δ could be defined as:

$$\Delta = \{x_y \mid x \in \{L, T, U\} \wedge y \in \{tr_z, fl_z\} \text{ where } z \in \mathbb{N}^+\} \tag{1}$$

and $\delta \subset \Delta$ represents a chronologically ordered set of the tasks we are using in our scenario.

We also define a binary operator ($\hookleftarrow$) and a function (*prec*) that we are going to apply on tasks

$$\forall x_1 \text{ and } x_2 \in \Delta, \text{ we write:}$$
$$x_2 \hookleftarrow x_1 \text{ iff } x_2 \text{ follows } x_1 \text{ directly or indirectly} \tag{2}$$

$$prec : \Delta \to \Delta, \text{ Where } x_2 = prec(x_1) \text{ iff } x_2 \text{ precedes } x_1 \text{ directly} \tag{3}$$

Hereafter, we may use annotations such as t_x, where t represents one of the time characteristics like CT, PDT, ST, and x is a task ($x \in \Delta$).

3.2 Predicting Violations Prior to Their Occurrences

The contribution of this section is twofold. First, it describes how to conduct predictive monitoring by mainly identifying events of interest, and second, it demonstrates how predicting violations prior to their occurrences could be accomplished.

Predictive Monitoring: To efficiently manage predictive monitoring, our solution consists first in automatically identifying the events of interest, i.e., the events to be monitored, and spare the user the tiresome work of defining them all at design time. The mechanism primary counts on the semantics of tasks. Although the low-level technical details regarding the semantic enrichment and analysis are still at an early stage, the concept yet deserves to be mentioned.

We distinguish between two types of events, *internal* and *external*. *Internal events* are the bits stemming from the process execution and may provide information related to the current location of the freight, indicating the start or end of a specific task, etc. They are mainly useful for run-time tracking. *External events* are the ones stemming from external data sources such as sensors, mobile devices, services, and IoT sources. They may provide information related to the process context where important knowledge can be extracted such as traffic congestion, airports security checks levels, etc. Internal events are easily collected from the process execution. However external events need to be identified based on the semantics of the process tasks. For instance, the status of all the planned routes between Maastricht and Brussels are considered as events of high interests for the task T_{tr_1}. Today, various initiatives relying on the crowd-sourcing techniques are delivering important contextual information such as Waze[2] for delivering road status. In order to collect such events, the *monitoring* component needs to subscribe to and unsubscribe from services delivering such events on appropriate times. For instance, if $ST_{T_{tr_1}}$ is equal to 8:00, the *monitoring* component needs to subscribe to the corresponding event source at least three hours prior to the task execution, i.e., at 5:00. Since in this paper we are not focusing on complex event processing, we associate specific static values capturing the delays that the process may come across. However delays stand as inputs to our system, and dynamic values are definitely supported. We suppose that for a **traffic congestion event**, 1 hour of delay is predicted, and for a **high security event** (at airports), 30 extra minutes are required. In this paper we discuss just the two aforementioned events for demonstration purposes.

Violations Prediction: Each time an event is monitored, we analyze its consequences by altering another time-aware representation of the tasks called the predicted process or graph (Fig. 3(b)). As this illustration shows, the first task finished successfully, but we suppose that while we were executing the second task (T_{tr_1}), we received the event indicating a traffic jam on the road causing one hour of delay. Subsequently The *Predict* component updates the predicted graph to grasp this new information, and to depict the expected future state of the process. To process the delay and sketch the predicted state of the process, the *Predict* component prolongs the impacted task $(T_{tr_1}$ in our case) while postponing the remaining tasks with respect to the underlying delay. Each task will start upon the completion of its previous one in order to maintain the logical flow. In other words, it is a delay propagation mechanism, where we continue propagating as long as the following task is flexible enough to accommodate the delay. Listing 1.1 stands as a slightly modified version of delay propagation algorithms used in tools from the project planning and management domains, like PERT. As one can see through Fig. 3(b), the task L_{fl_1} is postponed and it now ends at 13:00, however as the next task (T_{fl_1}) starts at 13:00, there is no need to propagate further. We see on the graph, that T_{tr_1} finishes at 11:00 and not 10:00, consuming the one-hour extra time that was previously accommodated between L_{fl_1} and T_{fl_1}. Sometimes we may not find enough flexibility to further propagate the delay, such as in the case of a flight transportation task. Thus, it is necessary to notify the user of a probable violation. This is formally defined in the following, where $Task_{aff}$ refers to the affected task:

$$\Omega = \{x \in \delta \mid x \hookleftarrow Task_{aff}\} \tag{4}$$

So Ω is the same as δ but without the tasks that precede $Task_{aff}$.

---

[2] https://www.waze.com/

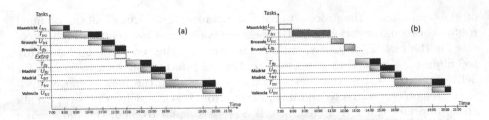

Fig. 3. Time-aware representation of the (a) Scenario (b) Predicted Process

Listing 1.1. The Predicting Algorithm

```
x₁ = PDT_Task_aff
PDT_x₁ = PDT_x₁ − delay;
if (PDT_x₁ < 0) Then predict a violation;
Prolong (x₁);
foreach x in Ω
  if (prec(x).ET > x.ST) {
  PDT_x = PDT_x − delay;
  if (PDT_x < 0) Then predict a violation;
  Delay (x);
  }
  else Quit foreach; # No Violation
end foreach
```

Revisiting our scenario discussed in Sect. 2.1. While we still on the road, we assume that we received an event alerting about the high security check at Brussels airport. Upon receiving such an event, the system will deduce a half an hour of delay to complete the task L_{fl_1}. Then the system attempts to update the predicted graph in Fig. 3(b). However as one can notice through the figure, $PDT_{L_{fl_1}} = 0$ (it has no black box attached to it) which leaves no room for any propagation. Then the algorithm will predict a violation, which triggers the *Recommend* component. It is important to highlight our technique that stands out from other current followed methodologies, as we are just checking the current ongoing process, but we employed complete predictive tactics, by trying to dynamically recover a future yet-to-come state of the process.

3.3 Proactive Mitigation of Predicted Violations

Once a violation is predicted, the *Recommend* takes the hand over. To deal with such a predicted violation, the component starts by searching for alternative solutions in order to comply to the Global Completion Time. The originality of our proposal lies in the attempt to adapt the smallest possible set of tasks.

The *Recommendation* phase defines the notion of sub-goals that characterizes a subset of atomic tasks. The first job of its algorithm, works on dividing the predicted process (Fig. 3(b)) into sub-goals. This is done by processing the tasks chronologically, beginning with the one that follows the current task and marking the Start Time (ST) of each subsequent T task, in addition to the Global Completion Time (GCT) as a sub-goal. From Fig. 3(b), we notice that we have three sub-goals ($ST_{T_{fl_1}}$, $ST_{T_{tr_2}}$, GCT). A sub-goal means a point in time that we need to respect, so our priority is to search for policies on how to protect the next sub-goal. Using this technique, our component seeks local adaptation means, before getting more and more global in scope. These local optimization strategies could be found in computing fields such as artificial intelligence, but they are not practiced in the business process management area. The Γ symbol stands for a chronologically ordered set of sub-goals. To build it, we apply this rule:

$$\Gamma = \{ST_T \mid \forall\, T \hookleftarrow Task_{curr}\} \cup \{GCT\} \tag{5}$$

where $Task_{curr}$ refers to the currently executing task. The next thing we need to build is the set of in-between tasks symbolized by Ψ. This set purely contains all the tasks between the current one (excluded) and the Next Sub-Goal (NSG). To build it, we apply this rule:

$$\Psi = \{x \mid \forall x \in \Omega \wedge x \hookleftarrow Task_{curr} \wedge ST_x < NSG\} \tag{6}$$

The following listing summarizes our proposed sub-goals algorithm followed with an explanation:

Listing 1.2. The Sub-Goals Algorithm

```
Build(Γ); # Divide the process into sub−goals
foreach NSG in Γ
 Ψ = buildSetOfInBetweenTasks(NSG);
  if (T ∉ Ψ) then solution = optimise(Ψ);
  else if (T ∈ Ψ) {
   first = Ψ.first; # Get the first task in the set Ψ
   solution = replan(first.ST, first.startLocation,
                     NSG − (loT + unloT), loc_NSG);
  }
if(solution ≠ null) then ProposeToUser(solution);
Quit foreach; # If user accepts the solution
end foreach
```

The loc_{NSG} variable represents the start location of the T task that has an $ST = NSG$, we can extract it from our annotated process (in our example, the loc_{NSG} for the first sub-goal is Brussels). Our main focus here is on the local adaptation strategy, and how business processes can be decomposed into sub-goals to help in readjustment. Although We completed the algorithm with two kinds of solutions that could be carried out (optimize and replan), yet the scope of applicable adaptations is wide, and studied in domains such as dynamic transport optimization and online re-scheduling. For more clarity, the difference between the optimize and replan are discussed below. The algorithm continues, to propose an alternative solution (if any) to the user, and then it quits if the user accepts, or else, the program loops again and tests the NSG.

Optimize vs. Replan: If there is no T task in Ψ, then the latter only includes L/U tasks. And to protect the NSG, it will be sufficient if we speed up these in-between tasks, in a way to execute them faster. Thus we search in involved companies directories for premium services that may eliminate the predicted violation, in order to propose them to the user. Executing the algorithm on our scenario, we will try to protect the first sub-goal ($ST_{T_{fl_1}}$). In this case, Ψ will contain U_{tr_1} and L_{fl_1}, so we will search if the first truck and flight companies offer some boosting services regarding the unload and load that they are going to do.

To illustrate the work of the replan method, we suppose that our algorithm will loop again and try to protect the next sub-goal ($ST_{T_{tr_2}}$). This time, $T_{fl_1} \in \Psi$, so our program will trigger a replan operation. This operation is well studied in fields such as dynamic and online planning, and it is not the main focus of our work. Yet we show that using the sub-goals notion we can algorithmically extract the needed parameters to help in *proactively* generating a plan that will keep the modification as minimum as possible. So the focus is not on the options we propose, because a whole set of solutions can be applied, but it is on the sub-goals decomposition, the automatic attribute extraction, and the predictive manner of engaging a solution. To continue the scenario, we assume that the planner finds a sub-plan that fits the requirement. This new plan demands to continue from Brussels by Truck to Antwerp airport, and then take the 13:30 flight to Madrid, to arrive there at 14:30. We propose the solution to the user along with some extra information to ask for her permission. After the acceptance of a solution we put it into action. For instance, we substitute the in-between tasks (Ψ) with the tasks of the new plan, and proactively handle a violation completely before it occurs.

4   Demonstration Prototype and Experimentation

Our components discussed in this paper have been fully implemented into a piece of software that is available online[3]. Interested readers can download it along with a video on how to use it. In the software we provided the user the ability to build his own logistics process model to simulate a plan (using the **Process Builder** graphical interface) or simply loading already stored process models. Among others, the user may specify the type of events simulation: automatic (the default setting) means that events are going to be randomly dispatched, whereas manual means that the user is in charge of generating the events. We will briefly explain the different use-cases we have generated for our experiments. Cases (a) and (b) correspond to the process used in the motivating scenario but the first is highly flexible while the second has no flexibility. Cases (c) and (d)

---

[3] https://dl.dropboxusercontent.com/u/13869335/pempra.html

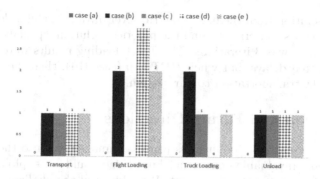

Fig. 4. The mode average of the number of iterations

characterize processes with a mono transport means, trucks can be delayed, but flights cannot. Finally, case (e) symbolizes a process that incorporates long operations across many connections.

One important dimension we studied was the number of iterations needed during the *Recommendation* phase (i.e., how many sub-goals need to be investigated to find a solution). If there is a single iteration in our **sub-goals algorithm**, this means that the program is proposing a solution to preserve the first sub-goal (i.e., the adaptation remains local). But if we end up with more iterations, the algorithm is trying to maintain many sub-goals (i.e., the adaptation spans over many more tasks). Our algorithm enters into the next iteration in two cases. First if there is no optimization or replanning solutions to maintain the specific sub-goal. But since we are not connecting to external services or real planners to check for solutions, this is not our interest for now, we are just simulating the availability of such solutions. Second, and more seriously, if the proposed solutions are impractical (i.e., they make no sense). In the latter case, the user will reject them and the algorithm will continue to the next sub-goal.

In our experiments, we notice that the appearance of impractical solutions essentially depends on the current task and when the predicted violation occurs. For instance in our scenario, if $Task_{curr} = L_{fl_1}$ and we receive a high security check event at Brussels airport, our algorithm will pretend that T_{fl_1} doesn't exist, so it will skip the first sub-goal and search for the second, and then it will propose solutions to boost U_{fl_1} & L_{tr_2} even though our system predicts that we may not catch the flight (fl_1) in the first place. Our algorithm fell short to handle such situations, because we are not supporting the adaptation of current executing tasks for now, this extension is part of our future plans. However if the events occur at different moments in time, the tool performs as shown in Fig. 4. This figure depicts the average number of iterations needed to find the first practical solution on all the different use-cases. The horizontal axis shows what was the current task when the violation was predicted. The average of case (a) was always 0 because of the flexible settings. As the figure also illustrates, if we predicted a violation during the transport and the unload tasks, no matter what the case is, the performance of the algorithm was masterly enough, as it proposed the practical solution from the first time. However during the flight loading task, the algorithm needs to iterate to find the practical solution, specially in case (d) where the average was 3 iterations. The Truck loading results were good when we have permitted delays, but when $PDT = 0$ (case (b)), then the settings will look like a flight transportation to our system.

5   Conclusion and Future Directions

In this paper, we carried the business process management into the next level, by showing how invaluable contextual knowledge could be exploited in order to create a safe execution environment. We addressed the challenge of predicting possible violations prior to their occurrences based on delay forecasts, then we explained how proactive adaptation is feasible. The work presented our proposed framework as an online prediction tool, and a proactive guardian for ongoing processes. To put such a framework into effect, a prediction algorithm that senses upcoming violations from context, along with an adaptor algorithm that proactively adjust and help in generating solutions, were introduced.

Integrating costs and penalties to our proposed adaptation actions, spanning to support multiple constraints, and fulfilling challenging multi-criteria (other than time constraints), are all lines of action on our future schedule.

References

1. Kephart, J., Chess, D.: The vision of autonomic computing. Computer **36**(1), 41–50 (2003)
2. Wang, C., Pazat, J.-L.: A two-phase online prediction approach for accurate and timely adaptation decision. In: SCC, pp. 218–225 (2012)
3. Leitner, P., Mechlmayr, A., Rosenberg, F., Dustdar, S.: Monitoring, prediction and prevention of SLA violations in composite services. In: ICWS 2010, pp. 396–376 (2010)
4. Metzger, A., Franklin, R., Engel, Y.: Predictive monitoring of heterogeneous service-oriented business networks: The transport and logistics case. In: SRII 2012 Global Conference
5. Engel, Y., Etzion, O., Feldman, Z.: A basic model for proactive event-driven computing. In: DEBS (2012)
6. Marrella, A., Russo, A., Mecella, M.: Planlets: automatically recovering dynamic processes in YAWL. In: Meersman, R., et al. (eds.) OTM 2012, Part I. LNCS, vol. 7565, pp. 268–286. Springer, Heidelberg (2012)
7. Maggi, F.M., Di Francescomarino, C., Dumas, M., Ghidini, C.: Predictive monitoring of business processes. In: Jarke, M., Mylopoulos, J., Quix, C., Rolland, C., Manolopoulos, Y., Mouratidis, H., Horkoff, J. (eds.) CAiSE 2014. LNCS, vol. 8484, pp. 457–472. Springer, Heidelberg (2014)
8. Cabanillas, C., Di Ciccio, C., Mendling, J., Baumgrass, A.: Predictive task monitoring for business processes. In: Sadiq, S., Soffer, P., Völzer, H. (eds.) BPM 2014. LNCS, vol. 8659, pp. 424–432. Springer, Heidelberg (2014)
9. Bassil, S., Keller, R.K., Kropf, P.G.: A workflow-oriented system architecture for the management of container transportation. In: Desel, J., Pernici, B., Weske, M. (eds.) BPM 2004. LNCS, vol. 3080, pp. 116–131. Springer, Heidelberg (2004)

Author Index

Aouadi, Hatem   69

Bizer, Christian   83
Bridge, Derek   17

Cremonesi, Paolo   45, 57

Deldjoo, Yashar   45

Elahi, Mehdi   45

Finance, Béatrice   127
Foping, Franclin   17

Guo, Bin   100

Jemaa, Maher Ben   69

Kaminskas, Marius   17
Khemakhem, Mouna Torjmen   69

Laender, Alberto H.F.   112

Meilicke, Christian   83
Meusel, Robert   83

Modica, Primo   57
Mousheimish, Raef   127

Pagano, Roberto   57
Paulheim, Heiko   30, 83
Pereira, Adriano C.M.   112
Peska, Ladislav   3
Primpeli, Anna   83

Quadrana, Massimo   45

Rabosio, Emanuele   57
Ristoski, Petar   30
Roche, Donogh   17

Santos, Hugo S.   112
Schuhmacher, Michael   30

Taher, Yehia   127
Tanca, Letizia   57

Vojtas, Peter   3

Zhou, Shasha   100

Printed in the United States
By Bookmasters